CRASH
TEST
GIRL

An Unlikely Experiment

in Using the Scientific Method

to Answer Life's Toughest Questions

HarperOne
An Imprint of HarperCollins*Publishers*

Praise for *Crash Test Girl*

"*Crash Test Girl* is a journey into the incredible mind and incredible life of Kari Byron, arguably one of the smartest women in the business. She is at once funny, profound, hilarious, and deep. This engaging and awesome book is the closest you'll get to a road map of how to be smart and savvy no matter what life throws at you. Strap on your seat belt and enjoy the ride!"

——Mayim Balik, star of *The Big Bang Theory* and
New York Times bestselling author of *Girling Up*

"When I was a kid I wanted to be Kari Byron. Actually, who am I kidding, I still want to be Kari Byron. This book taught me how."

——Simone Giertz, inventor, maker, robot enthusiast,
and YouTuber

"Supremely entertaining and surprisingly candid, *Crash Test Girl* is one part autobiography, one part live-your-life instruction manual presented with the unique badasscry that's made Kari Byron a total rock star and one of my favorite people."

——Mike Senese, executive editor of *Make:* magazine

"Kari Byron smashes the stereotype of a scientist as a white male in a lab coat with a beaker. Her fascinating journey from 'starving artist' to 'accidental scientist celebrity' gripped me at every page. Want to get more young women interested in STEM? Have them read this book!"

——Debbie Sterling, Founder and CEO of GoldieBlox

"Kari and I have been through a lot over the years on the set of *Mythbusters* and beyond. This book is like a behind-the-scenes look into the most epic explosions, gnarliest discoveries, and hilarious outtakes that happen when blowing sh*t up is in your job description."

——Tory Belleci, former *Mythbusters* host, professional builder,
and science entertainer

CRASH TEST GIRL

WRITTEN AND ILLUSTRATED BY

KARI BYRON

HarperOne

HarperCollins books may be purchased for educational, business, or sales promotional use. For information, please email the Special Markets Department at SPsales@harpercollins.com.

FIRST HARPERCOLLINS PAPERBACK EDITION PUBLISHED IN 2019

All photographs courtesy of the author

Designed by Janet Evans-Scanlon

Library of Congress Cataloging-in-Publication Data is available upon request.

ISBN 978-0-06-274976-5

19 20 21 22 23   LSC   10 9 8 7 6 5 4 3 2 1

FOR STELLA RUBY
AND ALL THE ADVENTURES
WE WILL CRASH TEST
TOGETHER

CONTENTS

- - - - - - -

INTRODUCTION

- - - - - - -

I f you're reading this, you probably know me as one of the science presenters from *MythBusters*.

Hate to break it to you: *I'm not a scientist.* Yep, that's right. I don't even have a background in science. No college degree in biology, or PhD in physics. No late nights in the library memorizing organic chemistry. No internships in a lab coat. In fact, the only times I've ever worn a lab coat were for photo shoots, when my co-hosts and I were gritting our teeth and trying not to roll our eyes. If you've seen *MythBusters* or any of the other shows I've hosted, this might surprise you. In fact, I've made a living from science entertainment, am best known for my contributions to the field, and even hosted the White House Science Fair and have inspired thousands of girls to become scientists themselves (or so they tell me via social media and at conferences).

One of the most gratifying things about being on *MythBusters* has been talking with teachers and reading articles that say our show changed the way science is taught throughout the country and abroad. Educators play our shows in their classrooms; parents rely on it for homeschooling. Kids re-create our experiments in their labs. They've used our methods as a jumping-off point to come up with their own myths to bust. When I was growing up, I hated my formal, textbook-y science class. Too much memorization, not enough inspiration. Now kids look forward to getting their hands dirty and using their imaginations in a less formal setting, thanks to the influence of our show.

We had no idea this sea change in education was going on while we were at the shop, working elbow deep in grease and grime. We were just regular people having fun and blowing stuff up. We didn't expect such amazing side effects, and we sure didn't expect to become a pop culture phenomenon. It all came as a pleasant surprise, especially for me. For many years, I was the only female host in science TV. At network parties, I would be the lone woman in sight. Now, there are many more of us on the air and in the room. I never set out to change things for teachers, parents, or other women in the field. But the status quo has changed, and I'm humbled and happy to have been a part of that.

So how'd a nonscientist like me end up with "science expert" as a job title? As it turns out, you don't need to wear a lab coat to be a science geek. You don't need a PhD to be fiercely curious. You don't need an engineering degree to figure out what happens when you put a Mentos in a Coke bottle. (Spoiler: it explodes!)

Let's throw out the old idea of what a scientist is. Say the word, and you might picture an old, gray-haired white man with glasses and a beaker filled with some mysterious substance. I was the opposite of that when I joined the show, a twenty-seven-year-old woman with an art degree and flaming-red hair, tattoos, and body piercings.

I was always taught that art and science were mutually exclusive. There are artsy types (me) and nerds (them), and I fell one hundred percent on one side of that line. But after a season on the show, I straddled it. **Art and science are both creative processes about using the tools at hand, being struck with bright ideas, and getting your hands dirty.** By living as a "scientist," I saw how adding an "artist" sensibility to science and vice versa lifted them both.

I started thinking of myself as something else, an experimenter. Our experiments weren't dry and precise, like measuring how much liquid to pour into a test tube. We were messy and bold. We were just winging it a lot of the time. And, while I may not be a "scientist" in the typical definition of the term, I do have a forklift license, keep duct tape in my purse, know how to blow up, cut in half, or light on fire basically anything you put in front of me, and have been conducting experiments professionally for *fifteen years*.

As a cohost on *MythBusters*, I wound up getting a world-class science education while conducting tests in order to prove (or disprove) all kinds of urban legends. Or as host Adam Savage used to say, "Helping settle bar bets professionally." Just as none of the hosts held degrees in science or were experts in those fields, the show was not originally intended to be a science show. But since we had to use engineering, chemistry, biology, statistics, physics, and math to, say, build a rig to split a car in half, we were thought of as a de facto science show after all. Our objective was never to teach or preach to viewers about the importance of STEM (science, technology, engineering, and math). We were just having fun and learning along with the audience about how things worked.

To do so, we employed the scientific method, the perfect narrative vehicle for proving and disproving myths. If you're not familiar with this elegant process for, well, figuring shit out, here it is in five simple steps:

1. **Question.** Learning begins with "I wonder what would happen if . . . ?" For example, "What would happen if I dropped a Mentos into a bottle of Coke?"

2. **Hypothesize.** Formulate a theory about what's going to happen based on what you know or *think* you know. For example, you could hypothesize that dropping a Mentos in a bottle of Coke would add minty freshness to the sugary sweetness.

3. **Experiment.** Do it. Drop the Mentos in the Coke and find out exactly what happens: Brown foam explodes all over the walls, the furniture, in your eyes, and up your nostrils.

4. **Analyze.** What the *hell* was that? Analyze the components of Coke and Mentos, study the chemical reaction that occurs when combined. Because the surface of Mentos is dimpled, dropping one in carbonated soda causes nucleation, the rapid multiplication of bubbles that have nowhere to go but up and out of the bottle. Hence, foam explosion.

5. **Conclude.** The answer to the original question is "If I drop this mint into the bottle of soda, I will need to steam clean the carpet." You should probably conclude to do it outside next time, with safety goggles.

Using the scientific method, my cohosts and I asked questions and tested over 900 hypotheses, filmed over 7,200 hours, set off 850 explosions, used 43,500 yards of duct tape, and loved every minute of it.

At some point during the show's long, fourteen-season run, it dawned on me that the scientific method could be applied to *everything*, and that I'd unwittingly been dropping the figurative Mentos into the Coke bottle of life *all along*, asking questions, forming hypotheses, experimenting, analyzing results, and drawing conclusions to figure things out.

Admittedly, my particular style of experimentation has been more like crash testing: I strap myself in, step on the gas, and head straight into a wall with my eyes open. It's always a crazy ride. I've gotten a bit battered and bruised (via bad decisions, bad relationships, dead-end jobs, to mention a few), but I don't mind my wounds. Scars are cool. Like wrinkles, they prove you've really *lived*.

If I hadn't crash tested through life, I wouldn't have landed the job at *MythBusters*, met my husband, made my best friendships, managed major depression, or gotten my wild life under control. I'm a living, limping example of how being a risk-taking methodologist can bring you love, success, and inspiration. **I had to be a crash test girl to evolve into a road-tested woman.**

If you're a professional scientist reading this, great! If you're a curious person who loves the idea of dissecting what went wrong and deciding how you might do things better next time, welcome! The brilliance of the scientific method is that it is for everyone. Seriously, *everyone*. The scientific method can become a guiding light for you, as it is for me, and I'm excited to show you how I did it.

4

The key is to fling yourself into experimentation and to be mindful about your results. My hypotheses were often misguided, as you'll see in each chapter ahead. My conclusions were usually unexpected. Wisdom lies in asking questions rather than in being presumptuous about knowing the answers. After things didn't go the way I thought they would a hundred times, I learned this overreaching Life Conclusion: **To be successful, you don't have to be right, but you do have to understand, with a scientist's emotional detachment, why you were wrong.**

With that in mind, I've written this book, a collection of stories about the many times I've been wrong, pushed boundaries and used critical thinking (How can this be better? What went wrong? What went right?) to understand what happened, learned and grew from it, and continually improved my life and relationships by measurable degrees.

I've organized this book into eleven chapters that describe how I've crash tested through the major areas of my life (not in order of importance!)—career, love, friendship, money, sexuality, depression, alcohol, setbacks, creativity, and bravery—and how I unknowingly was using the scientific method all along to ask questions, form hypotheses, experiment, analyze results, and reach conclusions. You can read the book from front to back, middle out, back to front, however you prefer. My message is to elevate experimentation and bring it from my life lab to your living room, and my goal is for the reading experience itself to be free-form and fun. I'm using every tool in my belt in these pages, and I've got a ton of them. Every chapter is packed with life hacks, rough drawings from my sketchbook, advice, experiments, lists and funny stories about my work on TV, real-life crash-and-burn moments, and all the times I've crashed and learned.

Before you start reading, I recommend fastening your seat belts. Helmets and safety goggles are optional.

CAREER

n *MythBusters*, we took on a "myth" that seemed so obviously true, I didn't understand why we were devoting screen time and energy toward confirming it. The Bull in a China Shop experiment premise was just a goof, a gag. We went to a rodeo bull pen, set up aisles of foamcore shelves (as not to hurt the bulls), and loaded them down with china platters, plates, bowls, cups, and let the bull into the pen to destroy it.

We were ready to film some beautiful high-speed footage of carnage. Instead, the bull trotted through the set and danced gracefully around the shelves without knocking over one dish. So we let in another bull. Same result. It was like they were dancing a pas de deux between the shelves, a graceful ballet of pirouettes, without breaking a single cup. We let in another, and another. Six bulls walked into the pen and up and down the aisles of shelves. In total, only one plate broke.

My cohosts Tory Belleci and Grant Imahara and I couldn't believe it. Tory broke more dishes while we were *shopping* for them. This experiment became very important in the overall arc of the series. A one-off gag that we thought was a "sure thing" taught us all a lesson (applicable for TV shows and life in general): **No one knows what's going to happen.** Even if you feel one-hundred-percent sure of an outcome. When it comes to assumptions, throw them out the window.

Finding My Dream Job

After college, I took a year to travel the world (more on that later). The same-age people I traveled with hoped to use their gap year to sort out some things before coming home and deciding on the next big step in their lives. One friend said, "I've figured out what I want to do. I'm going to grad school." Another said, "As soon as I get back, I'm going to get into public policy."

I ended up finding answers, but only to philosophical matters about who I was. Regarding my next step on the road of life, I had nothing. Well, I solidified in my mind that I was an artist, a maker. I knew how I wanted to approach life—as an introspective, mindful explorer with curiosity as my default setting. But as far as how I'd manifest this in the form of a career, I didn't have a clue.

I returned to my parents' house, a new place they had moved into while I was away, with my stuff in cardboard boxes. I was glad to see my parents, but it wasn't a happy homecoming. This wasn't home to me at all. I was twenty-three, I'd seen the world, and then I landed here, in a strange guest room in a strange house. I lasted a week before I decamped to stay with friends in San Francisco. I was broke, jobless, clueless, couch surfing. I hadn't found my life compass that pointed to my true north.

It was time to start over again and build a life from scratch. I didn't know where I was going, but I knew how I was going, with the spirit of adventure, and openness to experiment and try new things.

QUESTION

HOW DO YOU TURN YOUR PASSION INTO STEADY, RELIABLE INCOME?

How was I going to turn painting, sculpture, and bad poetry writing (aka my skill set) into gainful, gratifying employment, regular meals, and a decent apartment? Even free spirits had to pay their bills. In San Francisco, I'd be lucky to get four walls and a door with hinges instead of a beaded curtain. Since I had no step-by-step plan for how to begin an unconventional life, I turned to the conventional wisdom of beginning a traditional one:

1. Earn a college degree.

2. Get internships in your chosen field to build your résumé.

3. Step up to an entry-level job.

4. Upgrade to a better job at a better company every few years.

5. Land safely at the top with flexible hours, a hefty salary, and perky perks.

HYPOTHESIS

IF YOU FOLLOW THE TRADITIONAL PATH, YOU'LL EVENTUALLY LAND *THE JOB OF YOUR DREAMS*

A Job, Any Job

In the late '90s and early aughts, several years out of school, I was stuck at step two of the traditional five-step path I've described above, trying to get a handle on *what I was going to do with my life*. In the meantime, I had to earn a living, so I started temping all over San Francisco. I filed at a real estate office, fielded calls for a stockbroker in the financial district, catered galas for a food service company, and passed out flyers for a psychic on Market Street. A real psychic would know that NO ONE likes flyers.

I helped a pool shark at a bar distract his marks for a cut of his winnings. I even drove to wealthy communities to find designer clothes and shoes in the thrift stores, which I would buy for pennies,

9

and resell in San Francisco consignment stores on Haight Street for huge profit margins. (Seriously, a pair of barely worn Ferragamo heels for one dollar at Happy Dragon Thrift Store? Sold them for twenty dollars and I heard the buyers sniggering like they took me. Ha! Dinner and beer on me!)

I had a reputation for always having an angle or a game to keep my pockets full, but my dream job was definitely *not* side-hustling for dollars. The only dream I had was this: I wanted to somehow make money by making art.

I asked around for job titles with the word "art" in them, and that led me to dig out my best version of a secretary outfit and fake smile my way into a position at an advertising agency. It was downtown, two trains and a bus away from my apartment. This was before we had cell phones so I did a ton of reading on that commute. I would amuse myself by writing poems on playing cards and then leaving them for the next commuter to find.

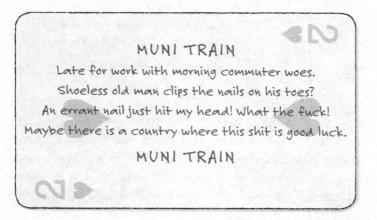

MUNI TRAIN

Late for work with morning commuter woes.
Shoeless old man clips the nails on his toes?
An errant nail just hit my head! What the fuck!
Maybe there is a country where this shit is good luck.

MUNI TRAIN

From my receptionist desk at the ad agency, I thought maybe I'd finagle my way into a lowly job in the art department and climb the ladder from assistant to art director. (See steps three and four from the "traditional path" list on page 9.) It seemed like a decent creative trajectory, accessible, and potentially lucrative. In the meantime, I answered phones. No one really talked to me unless you count "Wanna get a beer after work?," let alone mentored me. The phone I was tasked with answering barely even rang! All the calls came into the first floor receptionists. What they couldn't handle was bumped to the second floor, where I sat. This hardly ever happened. When the phone *did* ring, it was a thrill. I'd grab at it like a drowning woman would a life preserver for a brief interaction with a real human, and then it was back to the monotony of staring at my desk or doodling or occasionally looking at Myspace. (Remember Myspace? No? It was once a thing.)

I spent a year in this quiet desperation, sitting at the desk, staring at a phone. It became clear that I wasn't going to get a single promotion, let alone climb the corporate ladder to success and financial freedom. I began sending out résumés to other agencies, galleries, and museums. No one ever replied to my mailings. I'd fought to get my foot in the door, but I was just as stuck as ever, and my take-home pay was a lot less than reselling vintage clothes to thrift stores.

I realized something important: If I did somehow, magically, climb the ladder and end up with a job in the art department, where I'd be helping sell cars and cornflakes, would that be *fun*? **I'd been so focused on the path, I forgot to check where it led.**

Hot Shot

Meanwhile, as I continued to spend my days waiting by the non-ringing phone for someone to talk to or notice me, I signed on for a night job as a "sampler." Some people used the phrase "shot girl," like a beverage slave at a nightclub, but the spirits distributor I worked for considered itself too upscale for that description. I had no such illusions. You could call me Miss Piggy for fifty dollars per hour.

The job was to go to a few designated bars each night, give out two hundred coupons for free drinks, and talk up the cocktail with customers. I was trained on the process of making each spirit, and could answer almost any question about it. But mostly, I just danced away from bad pickup lines. If you were in San Francisco in the early 2000s, and a highly enthusiastic woman gave you free Scotch and waxed poetic on the finer notes of smoke or heather, it might have been me. I represented Tanqueray, Hennessy, Johnnie Walker, Grand Marnier, Moët, Cîroc, and several Diageo spirits.

While this job was a great improvement from my desk job, I always felt like I was missing something. "Sampling" wasn't creative or a positive contribution to society . . . it didn't have anything to do with art, or sculpture, or building, or creating—all of the things I was most passionate about doing with my life. But let me tell you the truth: I was really good at being a sampler. *Really* good. Doing anything half-assed is boring to me, so I put some gusto into the job. I have my dad's salesman's silver tongue and could talk to anyone. I played it like an actor, pretending to be carefree and excited to show up and dive head-first into whichever bar I found myself in. I even got a promotion! The boss asked me to manage a whole crew of sampler girls, dispatching them to bars all over town.

For a while, I drove to events in an Escalade wrapped with Tanqueray logos, which was super embarrassing. One night I was pulled over by a highway patrol and asked if I was drinking in that monstrosity. Seriously? Wouldn't that be a bit idiotic? Maybe he just wanted to investigate an SUV full of giggly girls in tight T-shirts.

Sampler-ing took me in and out of every nook and cranny in the Bay Area. I worked at a divey drag bar on Polk Street with a coworker who dressed head-to-toe in his best leathers. Then there was a backroom dance hall in Oakland where I was in the pale minority. At a bar in San Jose, I stood inside a cage and poured Hennessy into people's open mouths through the bars. I served Moët Champagne at the America's Cup yacht race when it was in San Francisco. At parties, food festivals, and Whiskies of the World events, I organized tastings, and was a spokesperson for Johnnie Walker and their umbrella-ed single malts.

The experience really drove home that people are people. You can talk to anyone as long as you're nonjudgmental and accepting of

WAYS TO KEEP YOURSELF ENTERTAINED AT YOUR ENTRY-LEVEL JOB

Not every first job is going to be as mind-numbingly boring as mine. But if you do have hours with nothing to do at work, try:

- **Doodling.** If I wasn't working, I was drawing. All of my pencils were extremely well sharpened.

- **Origami.** All those crafts you've been curious about? A boring desk job presents the perfect opportunity: knitting, embroidery, whatever. (My boyfriend got the most awkwardly long scarf ever for Christmas.)

- **Pro bono projects.** If someone was doing a charity event and needed poster art, I volunteered to make an illustration for it. I thought it'd help me advance, but those things turned out to be favors. People think art isn't work, and not worth rewarding. I made friends, but didn't get to the next level.

- **Reading Missed Connections on Craigslist.** People wrote two or three of them about me. One had collected my poetry postcards on the 22 Fillmore bus and wanted to meet me to discuss them. Maybe he'd found a chicken foot under his seat, too, and wanted to compare notes. I never replied to any of them. I was a stealth poet. I liked to be mysterious. While traveling, after I finished a book, I wrote a note in it and left it in the hostel. I leave notes in hotel room drawers. I just love the idea of connecting anonymously with strangers and never knowing the outcome.

- **Updating your résumé.** Nothing like getting a paycheck and using the resources at the office to search for another job.

13

whatever personality is in front of you. This skill came in handy later on at *MythBusters* because we employed experts in a wide range of different fields. Even if I didn't speak their particular lexicon, I could still converse with them without feeling intimidated by them or embarrassed by the gaps in my education or experience. Whether they were Civil War reenactors, biology professors, or arms dealers, everyone has a story. I loved exploring the anthropologic beauty of everyone who came to our shop. (Having a vast knowledge of alcohol also helps in dealing with *MythBusters'* Australian film crew.)

My job made me very popular, especially when I was assigned to be a secret martini taster. I would go into a bar, and ask the bartender to make the brand's signature cocktail, taste it, and ask him or her about the brand and the drink. Then I'd write a report about it. I would go to three places a night, and go through the same process. As a petite woman, I couldn't possibly finish all the drinks I had to order and taste. While I took notes on how the bartenders made the drinks, I had some "designated drunk" volunteers—my broke artist and musician friends—who would help me finish my assignment. Best job ever for living a *Sex and the City* life on a *Shameless* budget. (I watch too much TV.)

I was enjoying myself, and the bosses could tell. They gave me more and more responsibility, and raises to accompany it. For the first time in my twenty-something years of life, I made enough money to pay my rent *and* buy groceries.

But something nagged at me. All of the other sampler girls saw this as a temporary gig, a way to get paid while continuing their educations—and most of them did go on to have professional careers as doctors, lawyers, and engineers. But I had no plan B, no alternative. I was living for today. Was this the track I wanted to stay on forever?

I reminded myself that when I dreamed of being a maker, it wasn't a maker of cocktails. I was a great liquor ambassador, had fun, and basked in the appreciation and acknowledgment of my bosses. But I had the sense that I should be doing something else, there was something else out there for me. However, like almost everyone who finds themselves in this predicament, I didn't know where or how to start over. How much longer should I keep hoping for a creative career when I hadn't gotten close to one yet? Were my passions just pie-in-the-sky fantasies? Or were they something I should chase in the form of gainful employment . . . ?

I asked myself this question constantly during my excruciating, limbo/first-circle-of-Hell hours at the ad agency job (I hadn't given that up; I was receptionist by day, shot girl at night). To keep myself from going crazy, I started bringing polymer clay—the stuff special effects artists and prop makers use for toys or models—to the office, made sculptures at my desk, and amassed a large collection of little statues. I showed them to a friend of mine who was trying to find his way into the movie industry.

"You should do *this*," Matt said of my models. "You're great at it. Put together a portfolio."

"No, no, I'm not good enough," I said.

I was afraid of being judged. A real professional had never seen my sculptures. What would a *real* artist think? Would he or she know I had no clue what I was doing?

Now, when I look back on this time in my life, I'm aware of my inferiority mind-set, always seeing the world from the outside in, a stubborn holdover from junior high. Now I see the world from the inside out. This is my art, like it or not. I do it for me. If you like it, great. If not, great. It's not up to me to decide your taste for you.

Matt kept on encouraging me. One night, he said, "Listen. There's a local place called M5 Industries run by this guy Jamie Hyneman." He described it as a design and build company/workshop that made special effects, prototypes, and props for commercials, movies, and TV. "Jamie's assistant Alberto teaches a class out of the shop on sculptural special effects and model making," he said. "Why don't you come with me and just check it out? You will love it."

And *still* I hesitated.

15

Sculpture was my favorite thing to do in the whole world. Was I really going to give up on my dream so quickly? Was I really going to *assume* that this job would be like the others, that I was never going to make things for a living? It was time for me to throw out my expectations and my baggage, but, most importantly, my assumptions. About myself, about life, and about who can and can't have the career of their dreams. I searched my mind for a mantra. At the top of the list was the all-purpose powerhouse: "Why *not*?"

I reached into my soul and told myself, *Woman up, Kari! Matt is right. Yeah, special effects! You could do sculptures for a living!* I forced myself to push past fear, and go for it.

**TRY ANYTHING,
DO EVERYTHING,
SAY "YES" EVEN WHEN
YOU AREN'T SURE**

EXPERIMENT

Beg Your Way In

The next day, as I sat at the second-floor ad agency reception desk, I put together a portfolio with every drawing and sculpture I'd made at my job in purgatory. I was nervous about the possibility of another disappointment at M5. I'd already sent résumés with no response to every animation and special effects shop I knew: Pixar, Industrial Light & Magic, and others. From my lonely desk on the second floor, I imagined my résumé and portfolio being tossed over the shoulder of an eye-rolling HR person into a comically large bin labeled REJECTS. The beautiful outcome of feeling like a reject for so long: **Desperation makes you bold.** And that boldness led me into the greatest opportunity of my life.

I went to the class Matt recommended, and I walked into M5 Industries for the first time. My memories are always a bit rosy, but I'm pretty sure there were angels singing and God-light emanating from the shop door. There was a metal shop, a wood shop, an electronics shop, a mold room, a paint room, movie props all over the walls, cool sculptures (including a Mugwump from *Naked Lunch* in the corner), spaceships from *Star Wars*, masks and models from *The Nightmare Before Christmas* on the walls, and the shuttle from *Running Man* with a half-size Schwarzenegger hanging from the ceiling! I can still feel my excitement. (I am even typing this section fast remembering this moment.) There was every tool you could imagine, and boxes upon boxes with labels like GOOGLY EYES, PLASTIC DOMES, TANK PARTS. I could smell freshly cut wood, metal dust, and the slight scent of toxic fumes from resins and paints. To someone like me, it was like new car smell or freshly baked cookies. All of a sudden a light went on and I thought, *Oh my God. I want to work HERE!* **I had to be a part of this place. I'd do anything to stay.**

I went up and introduced myself to Jamie Hyneman. He was wearing his signature beret and big moustache, a look he'd maintained for decades. I was wearing my best "don't care but I actually really care a lot" outfit (jeans and a black T-shirt, combat boots, and a flannel tied around my waist; grungy but not dirty). I tried to seem full of confidence but my goofy grin and nervous talking didn't help. Jamie is an intimidating character; he doesn't smile needlessly and definitely doesn't suffer fools. I was used to people beaming at me from my night job (offering free drinks will do that), so I had no idea how to gauge Jamie's reaction to me. I learned that he's incapable of lying to soften a blow. He is straight with you, a trait that I both respect and fear to this day.

I showed him my portfolio. He flipped through, page by page, with silent disinterest. I swallowed back nerves and made an audible choking sound because my mouth had gone dry. With every page he passed, I died a bit inside. When he got to the very last page, he pointed at the picture of a wrinkly old man sculpture I was pretty proud of, and said, "Maybe we can work with that." Exhale.

"Well, I'm willing to intern for free," I said.

"Come back tomorrow."

17

TOP THREE GET-OUT-OF-WORK-FREE CARDS

1. **Pink eye.** My favorite ailment for getting out of work, it is disgusting, highly contagious, cleared up in a few days, and no one wants you within five miles.

2. **Food poisoning.** The gross factor is high, and most prefer the symptoms not be described in detail. You have to stay home long enough to make sure it isn't the flu or Norovirus, so it's an excuse with an open end date.

3. **Family emergency.** It sounds dire but is purposefully vague. No one would be so rude as to pry what the emergency is. And if anyone does, say, "My mother has food poisoning, the flu, the Norovirus. And pink eye."

And that was it. My foot was in the door again, but this time, in the right place. I was thrilled! I was elated!

But, crap. I still had a nine-to-five job at the ad agency. Should I call in with the flu or pink eye? I didn't know at the time that I would never go back to the silent phone on the second floor. I would never go back to being a corporate zombie. Instead, I just kept showing up at M5, every tomorrow, made prototypes and toy parts, learned from Jamie, and loved every minute of it.

I was so energized about doing something art related at M5 by day, the good vibes spilled into my night job. Out of the blue, one of the spirits companies offered me a big job as a brand ambassador with a significant salary, benefits, perks, and a future. My job would be going to dinners and events, like I'd been doing, plus I would travel to exotic places representing the brand and running promotions. I could see a career and a future taking shape in front of me, making more money than I had ever imagined.

So, what was it going to be? The high-paying, high-flying corporate job I could grow into, or the unpaid internship that could end at any second?

I was given two days to decide.

It took me two seconds.

Everyone in my life thought I was crazy—except Paul, my then-boyfriend, now-husband, a fellow artist dreamer weirdo. As my friends reminded me, "You're not a kid, Kari. You can't play with clay and make toys forever. What about earning good money and having security and maybe growing up?"

My parents asked, "Are you sure?"

I told Mom, "The risky path is way more interesting. I'm gonna go with that."

I was poor but happy. It felt like a new adventure was beginning, and was it ever. My decision turned out in my favor, I have to say. If you are reading this book because you're a fan of *MythBusters*, you already know that.

YOUR INSTINCTS KNOW WHICH JOB IS DREAM-WORTHY AND WHICH IS A SAFE COMPROMISE. TRUST.

ANALYSIS

Use Everything You've Got

My first day as an official intern was, coincidentally, the first day *MythBusters* was being filmed at M5. I had no idea that it was going to be the set of a TV show, and, after I found out, I certainly had no expectation that I'd be a part of it. There were other interns and shop guys hanging around. I was just one of the crew, just glad to be there.

It was 2002, the Wild West of cable, before reality television was highly produced, contrived, and scripted, spread out over dozens of networks. Discovery Channel had always been known for nature documentaries, but the corporate overlords decided to try something new and made the shows *Deadliest Catch*, *American Chopper*, *Dirty Jobs,* and *MythBusters*. They were unfamiliar departures from the norm, and successfully ushered in a new era of content. *MythBusters* was set in a working shop, and anyone might get pulled into the action—or, in my case, be enlisted to sweep up the action after the camera was put away.

One day, Jamie and Adam (the hosts), and the show's creator, Peter Rees, were planning the segment called Vacuum Toilet, about the legendary myth of a very girthy woman sitting down on the toilet in an airplane, flushing it while seated, and creating such a powerful suction around her fleshy rear that she got stuck on the seat until the plane landed.

Jamie called me into his office and said in an awkward monotone/ half giggle, "So, Kari. In exchange for learning how to use the software to create a mold, would you let us do a 3D scan of your backside? Because we need a female butt." (The way Jamie enunciated the word "butttttt" with five extra t's still makes me smile. He was probably just as uncomfortable asking me this as he knew I would be hearing it.)

My butt is undoubtedly female. I was also the only girl hanging out at the shop. But make a mold of it? I wasn't sure about that.

"And I will give you a hundred dollars," he added.

"YES!"

Besides, who was ever going to see this? I mean, it was just a weird cable show.

Also, no one had a clue on how we were going to actually make the mold. The beating heart of *MythBusters* was coming up with solutions to ridiculous problems. We just strapped on our thinking helmets and figured things out.

The production manager went shopping and bought me this ugly flesh-colored one-piece catsuit to wear while they did a 3D scan using super high tech that is now antiquated software. I leaned over a stool, held still for half an hour for the scan, and had a suspicion that the male interns were taking pictures of my bum.

If you are wondering right now whether I objectified myself, a betrayal of the feminist cause, hold on for a second. During my sampler-ing days, people stared at my butt every night—like every other girl in the bar. It wasn't part of the job, but it was an unavoidable hazard of it. In this case, doing the 3D scan of my butt was an ass-ignment (Yes! I am the pun queen.) I was uniquely qualified for in a shop full of all men. It just didn't feel exploitive. *It was for science.* And one hundred dollars. Somehow I knew that this odd opportunity could be my ticket to more.

Where does that land me on the feminist scale? My attitude was (and is) if you give me a wrench, a hammer, and a saw, I'm going to use every tool in the box to get what I want. In this case, I wanted in on the project. They needed a female butt. I happened to have one right on me—the only person who did—so I used it. If a guy gets a job because of his physique, he's not exploited, but if I used my shape to get this one, it is? I was interning while female. My boss needed a woman's rear. If I didn't use that tool, I would have neutralized myself out of an opportunity and that was not going to happen. In my years of job shifting, I'd learned to play every angle and use every attribute I had. I was only being consistent. **Always say "yes" to opportunity. You can change your mind later. But if you say "no" at first, you might not get another chance.**

After the image was downloaded onto the computer, I used the sculptural software to design and refine the 3D model. My actual butt wasn't big enough, or cushy enough, to test the myth. I was tasked with the artistry of making an enlarged version of it look "real," so, as a reference, I went online and looked at a lot of plus-size porn. It was the only way to see butts, perfectly dimpled, for my rendering. Everyone that walked past my computer and saw what I was looking at just nodded and walked on. It was that kind of place. As per our deal, I learned the new tech and got my hundred dollars. Jamie got a 3D model of my butt on his hard drive. Hell of a deal if you ask me.

As far as my original belief that no one would ever see it, the video clip of my catsuit scan has been watched five million times on YouTube. On the bright side, it was my twenty-eight-year-old butt. Who wouldn't want a record of that?

That was how I got my butt on TV but not how I proved myself an asset. (That pun was too easy.) What really showed my dedication was my persistence during the JATO rocket car myth. After a full day at M5 and a shift at my night job, I jumped in my old Toyota pickup to drive all night long to the Mojave Desert and meet the crew by morning. I had a Hunter S. Thompson spoken-word tape stuck in the deck and was almost driven to madness but I wasn't going to miss that old Impala screeching across the desert landscape propelled by three actual JATO rockets! I got to the hotel parking lot by 4:00 a.m. and slept in my truck for an hour. When the film crew caravan headlights hit my window, I jumped out of the truck and ran to catch them. Everyone was looking at me like I was a crazy person. They laughed and said, "What are you doing here?" "I came to help!" My stalker-ish insistence to help out worked. I made myself valuable by always going the extra mile, or in this case, five hundred miles. It was so worth it!

The first season of the show was only three myth-packed episodes of mayhem. I found myself helping more and more. By season two, the production team realized they needed to speed up the process of making episodes. They asked if I would become part of a team that would help do all the building, setup, and cleanup for Jamie and Adam so that they could crank through the filming faster. Our work would be behind the scenes only. Since I loved helping on the show and there was a promise of a small salary, I jumped on that.

My butt got me in the door but this day got me respect.
After my night job, I drove all night to the Mojave Desert.
I slept in my car for two hours in their hotel parking lot
and waited for the headlights rolling out to location at
5:00 a.m. I flagged them down and spent the day assisting
in whatever they needed. I busted ass to help the JATO
rocket car myth for the first episode of **MythBusters**.

Building a Team

I met Tory Belleci in the parking lot on his first day as a builder on the show. He was rolling a huge tool chest toward the shop.

"Hi. I'm Kari," I said. "I think I'm going to be working with you."

"I guess you're going to be my right hand," he said.

I'm pretty sure I giggled nervously before saying, "I could be your left, too."

What does that even *mean*?! We still laugh at my awkward eagerness.

Tory had just come from Industrial Light & Magic, George Lucas's special effects shop, working on projects like *The Matrix* and the new *Star Wars*. Basically, I wanted to be him when I grew up. He had the fancy special effects job that I thought was The Coolest, so I was hungry to hear all of his stories. We got along really well, mostly because when you're in somewhat miserable circumstances, you tend to bond with people, and we would often find ourselves at M5 until 10:00 p.m., scraping chicken skin and guts off the ceiling after a long day testing the Chicken Cannon Myth, for example. With blood and meat bits and the sweet smell of salmonella on his hands, Tory asked, "*Why* did I leave ILM?"

"I don't know, man. I don't know," I said. I was so glad he did.

After several weeks, the production team realized that the new format still didn't pick up the pace fast enough and they needed to change it again. "So, you're hosts now," they told us. "We need you to start talking to the camera." That meant Tory and I, Scottie Chapman (another local builder/rock star welder), and a "myth-tern" contest winner named Christine Chamberlain were now going to be on-camera talent?

I said, "So other people are going to actually *see us*?" None of us was an actor; none of us had any *ambition* to be an actor. We were just background builders, promoted to being hosts, being thrown in front of the camera!

It was a christening by fire. When you look at those early episodes, it is *super* awkward watching us talk on camera. I was terrible at it at first.

The crew of *MythBusters*, from the beginning, put in thirteen-hour days, six days a week. We traveled together, and worked side by side. We might've started out as strangers, but we became family fast. And, like family, sometimes we loved each other, sometimes we hated each other, sometimes we got on each other's nerves. But we were stuck together through good, bad, and ugly times.

After a year or two, Scottie and Christine left the show and we needed to find a new person for the build team. Tory and I said, "Grant, of course."

Grant Imahara was always coming into Jamie's shop to help out with electrical things, like remote controlling, electronics, and puppet props that were used in commercials. When I met him, Grant was just a sweet, scary-smart, *Star Wars*–obsessed, adorable, friendly uber-nerd. It looked like he was trying to grow a mustache that never really got anywhere, which was endearing.

We interviewed other people, but that process only confirmed our belief that Grant was the guy. The network wasn't sold on him, though. Television people will always push for a more "polished" human, which was absurd in our case. We were supposed to be builders, not supermodels.

Finally the network gave him the okay—with the caveat that he needed a makeover. They sent over pictures of Brad Pitt and Colin Farrell and said, "Like this." The show creator Peter Rees pushed back and said, "We're not changing a thing. You hire the super nerd, you get the super nerd." Our being just regular people was the appeal of the show. None of us were brought in to be well spoken or beautiful. We were just us.

The network realized our appeal was that we were real—everyone knows a Kari or a Tory or a Grant. The show kept getting renewed,

and we were more and more inventive and daring as the years rolled along. I remember coming home to Paul and telling him, "I shot a propane tank with a rifle and blew it up. It was the coolest day!"

He said, "This is not your day job, this is your dream job."

THE HARDER I WORK, THE LUCKIER I GET

CONCLUSION

My path from receptionist/cocktail pimp to builder/TV star was like an Escher floating spiral staircase in the clouds. I was never sure if I was going up or down, if it would ever end, if I'd have to lean over the side and puke, or if it would turn into a slide and I'd fall into the sky. At no point until I sat down to write this chapter did I ever stop and think, *Well, Kari, you sweat, risked, begged, and bombed your way into a brilliant career!* I didn't conceive a solid idea of success until I already had it. I assumed my career would go one way, and it turned out to be something I couldn't have possibly imagined.

Any creative endeavor is like that Bull in a China Shop experiment. You have to keep an open mind. Go ahead and make your assumptions and hypotheses. It can't hurt to form a working theory based on what you've heard, but don't stubbornly cling to it, or you could miss out on the wonder and wisdom being absolutely, fantastically clueless and wrong. It happens all the time: Someone thinks they know more than they actually do based on an assumption. But assumptions, even seemingly safe ones, aren't proven as fact unless they've been tested.

It's wiser to open yourself to possibility and dumb luck, let them do their thing while you get out of your own way, avoid traps, take risks, and bust (mold, whatever) your ass.

And don't give up. At twenty-eight, thirty-eight, or eighty-eight.

The only thing I knew starting out on my career path was that I wanted to make stuff. I wanted to get my hands dirty. I thought it'd be using paint and clay, but I wound up using metal, chemicals, and explosives. I hoped that I'd get to work with people I liked and respected, and I got Tory and Grant, colleagues and lifelong best friends.

In the final analysis, my original hypothesis about how to find a dream job wasn't so far off, but only in the broadest terms:

I got an internship.

The internship turned into an entry-level job.

I got promoted into a better job.

It became a bug-eating, poop-collecting, chicken-exploding career.

28

LOVE

Early on in *MythBusters*, we did an experiment about whether talking to plants made them grow larger. My first reaction was that it seemed like a lot of crystal-gripping hippie nonsense. But we set up different green-houses of vegetable plants—one with plants we said nice things to ("I love you, you're beautiful, plant"), one with plants we were mean to ("I hate you, you're disgusting, plant"), another that received a steady onslaught of Swedish death metal music (you get the idea), and the control group of peace-and-quiet plants that we ignored. All the test subjects received the same sunlight and water.

Much to my surprise, the plants that we talked to, and especially the ones that rocked out to music, did grow stronger and faster than the ones that we ignored. Despite my eye-rolling and general disgust about saying, "You're the best plant ever!" in a sickeningly sweet voice for days on end, it turned out that there was a scientific basis for our results. The aspiration of our breath—basically, blasting CO_2 at the plants—and the vibrations of our voices and the music left micro-scopic breakage in the cell walls. In sustaining minor damage, the plants rapidly repaired themselves and became bigger and stronger, just like a healed bone is thicker at the site of the break.

By applying all-important critical thinking, we learned that plants like to be yelled at, and that not talking to them left them limp. I also learned that the entire cast of *MythBusters* had brown thumbs. We didn't do many experiments with plants after that because we knew we'd end up killing them all.

Welcome to My Wonder Years

On the first day of the third grade, the world became bigger than teddy bears and LEGOs. Enter the bad boy. He had bright red hair, freckles, a mouth full of braces, and acid-washed jeans—the mark of a fashion rebel, even for an eight-year-old. I also found myself curiously fond of a smart, funny kid with baby-fat cheeks and a quick wit that kept the teachers on their toes with a timely fart joke. I used to listen to Michael Jackson's "The Girl Is Mine" and imagine the two of them fighting over me. (You know those things you never tell anyone and then you find yourself writing in your book? This is one of them.)

Bad Boy vs. Funny Guy has been a recurring theme for me throughout elementary school, and well into junior high. I was torn between *21 Jump Street*–era Johnny Depp and *Splash*-era Tom Hanks. The one thing the redheaded bad boy and the baby-cheeked funny guy in my third grade class could agree on? Neither wanted anything to do with me. (Unless you count gluing my ponytail to the desk behind me! I will never totally understand taunting as a sign of affection.)

At fourteen, I was a shy girl, and could barely speak to a boy, much less attract one and then put my tongue in his mouth. My secret crushes made me break out in a flop sweat in real life, but they played starring roles in my diaries where I pined poetically for them. I could fill pages with imaginary trysts, but when I saw "him" (whomever "he" was at the time) in the hallway, I'd look at my feet, and die a little inside.

According to the junior high social hierarchy, girls with boyfriends were popular or rich, soccer or track stars—with boobs. I was a late bloomer, as well as shy, definitely not rich, nonathletic, skinny, and incapable of speech. But, as I learned in the movies, girls didn't have to make the first salvo. "The boy should talk first," I told myself to let myself off the hook. I remember walking my dog by this parking lot where a bunch of skater boys hung out. My secret tactic to get them to notice me was to furiously and passionately ignore them. Not once did any of them chase after me, ask me my name, or even say, "Hello."

Didn't they get that I acted like they didn't exist in order to get their attention?

Why were boys so *dumb*??

QUESTION

WHAT IS TRUE LOVE, WHERE DO I FIND IT, AND HOW DO I MAKE IT LAST?

First Contact

Jon was a trumpet player in the marching band, and I thought he was super hot. When he walked by, my girlfriends and I suddenly became fidgety, quiet, serious, keenly interested in our books. Couldn't show that we noticed him, or that would break the illusion that we didn't give a fuck, which was the only way to show a boy that you absolutely DID give a fuck.

According to all of human history and John Hughes movies, the trumpet player in the high school marching band was, by definition, a geek. Jon was the world's sexiest geek then. I didn't mind. I was a geek, too. I was an ID girl, which meant that, at halftime during football games, I twirled a flag, and in parades, I held a shield along with eight other girls that spelled out the school name. Mine was the space between L-O-S and G-A-T-O-S.

I joined band for a few reasons: 1) It came with an automatic friend group, 2) I was excused from gym, and therefore super-shy me didn't have to shower with girls who all seemed to have developed into actual women while I still went braless, and 3) I got to wear a glittery

SOME THOUGHTS
ON MARRIAGE

My early crushes on Tom Hanks and Johnny Depp were filled with fervor. But I never fantasized about marrying them, not even Tom (I know, I know . . . he's basically the epitome of "husband material"). I didn't realize until much later that this was odd. Marriage never crossed my mind, and I never associated myself with the role of "the wife." Instead, I was in love with pure romance: raw, passionate, dramatic—like a horrible rock ballad. But also like the love I saw at home. My own parents were soul mates—still are—after forty-five years of faithful marriage.

I liked to play with Wedding Barbie in her pretty dress—who didn't?—and my version of Ken clearly understood that Barbie had her own friends, her own pursuits, her own career. Even as a girl, I envisioned love with lots of personal space and freedom. If I wanted to up and travel for a month, my partner would have to say, "Can I help you pack?"

I also thought of myself as a romantic, in love with the heroines in *Breakfast at Tiffany's* and *Gilda*, naively seeing them as free-spirited women who went to parties, seduced handsome men, were strong, beautiful, and fascinating, *and* unafraid to show their vulnerability.

My desires for independence and freedom seemed to contradict my equal longing for true love and romance. I wanted nothing and everything from a relationship, and I knew that finding a partner capable of giving me what I wanted was not going to be easy. It might not happen at all.

costume that I could pretend to make fun of but actually loved. In terms of high school hierarchy, the marching band was low on the totem pole. And in terms of marching band hierarchy, I was very much at the bottom. But Jon was a trumpet player and a senior so he was quite a catch.

The band traveled to competitions in a bus, and I always wanted to be within a few rows of Jon. That way, I could employ my "ignore him and never make eye contact" strategy. But Jon was nearly as shy as I was, so I switched my game to "have my friend tell his friend that I thought he was cute."

It worked.

LOVE ONLY WORKS WHEN YOU WORK FOR IT

HYPOTHESIS

One night, he walked me to my front doorstep. I knew what was coming and was terrified because I had no idea what I was doing. We leaned toward each other, and I got my first *Princess Diaries*–style one-foot-popping kiss. It was awesome.

After the sweetest first kiss that ever was, I said, "Good night," ran inside, closed the door, ran to my bedroom, and called my best friend Brittany to tell her every tiny detail. Being "with" Jon raised my social status by a mile. Suddenly, I was part of a group of girls who traveled in a clump in the general proximity of the clump of our boyfriends. I can't remember going on a one-on-one date with Jon. Maybe he never asked, or maybe I avoided it. He might try to run the bases with me, and I wasn't ready to do that. Or was I? What if he touched under my shirt and I hyperventilated? That'd be so sexy.

It never got that far anyway. It was the innocent high school era of dry humping on the couch while watching rental videos—the '90s version of "Netflix and chill." We were solidly planted on French-kissing first base, nearly always surrounded by other people. As a couple, we barely talked and never removed clothes. I was very satisfied with this level of intimacy (hardly any). I was far more excited to have girl friends who participated in my "romance" by sharing our stories and secrets. **I enjoyed being in a couple just as much as I liked my boyfriend.**

THE FIXER-UPPER

CRASH TEST

When I was eighteen, I fell for a guy named Jason, like the horror movie. He had an infectious laugh, like a machine gun, which I adored. Jason was the first person I said "I love you" to. We were together for a year before this sweet, ambitious, funny guy took acid, and he didn't come back from the trip. He started speaking in numbers, spouting conspiracies, and went crazy with paranoia. He ran from his house in the middle of the night and showed up at my window, soaking wet, no shoes, ranting with wild eyes. I stopped him from burning his house down and from cutting his face with a razor. I was afraid to leave him alone.

His paranoia became so stressful for me to manage that I got really sick, too, and my parents wouldn't let me leave the house. My friends encouraged me to break up with Jason, but I stayed through all the craziness because I felt it was my duty to help him recover, which he eventually did. I thought the strong thing to do was to fix him/us. I thought that love was something you worked for, worked at.

A classic case of "no good deed goes unpunished," soon after he recovered, he started cheating on me with an actress. One day I found a practically naked picture of her in his drawer and asked, "What the hell is this?!" Looking back now, I think he wanted me to find it. The conversation he wanted to have with me wasn't one he could just bring up casually.

"I've been seeing someone else," he said. "She's pregnant and I'm going to move in with her." Within the week, he packed all his stuff into a U-Haul, and ran off to Texas to be with his new girl.

The day he left, I went over to his apartment to say goodbye. I pretended to be cool with his new life/family, and told him, "It's okay. No hard feelings *at all*. Take these as a peace offering."

I had baked him a batch of cookies laced with saltpeter. My friend told me it would make him flaccid, and I needed revenge. Don't try this at home, readers. I can't recommend that anyone poison their ex (although, this guy had it coming).

Lil Q **What Is Saltpeter, and Does It Really Make a Dude's Dick Limp?**

Saltpeter is the compound potassium nitrate, used for many practical purposes, like oxidizing black powder in fireworks, salting meat, and thickening soups and stews. It's a legitimate plant fertilizer, tree stump decomposer, desensitizing ingredient in toothpaste, and a treatment for asthma and high blood pressure.

Somehow, it got a reputation as an anaphrodisiac, a turn-off drug, and is believed to have been added to the breakfasts of prisoners and military personnel to decrease their sex drive or make them impotent. Effective? Probably not. If anything, the saltpeter made them so sick to their stomachs that the last thing they wanted to do was have sex.

FACE OFF

On the show, we did an episode that pitted men and women against each other in their ability to read facial cues. To test which gender could read them more accurately and faster, we took photographs of our faces making seventeen basic emotions—happy, sad, scared, angry, confused, etc.—and then blocked out the photo to show only the eyes. We asked a dozen women and a dozen men to identify the emotion revealed in just the eyes before showing them the entire face.

The results? The women were significantly better at identifying the emotions accurately, and way faster to call them out. The men dithered and made what were plainly confused faces before they gave up answers.

One weird trend: For whatever reason, my angry eyes were mistakenly identified by women as "happy," and by men as "sexy." Grant's angry eyes were called "come hither" and "flirty" by a lot of the women. It would have been useful to know when I was single that glaring at men would turn them on. On second thought, I think I had figured that out . . .

It was months, maybe even a year, later, and I was hanging out at a party where people were swapping urban legends about ex revenge stories. Like the one about the woman who spread grass seeds over her boyfriend's living room and flooded it while he was out of town. It always happened to a friend of a friend of a coworker. My face almost cramped up and I had to bite a dent into my tongue trying not to laugh as a dude told the saltpeter story. Funny to think I had a future in busting legends like these since I had clearly contributed to the quilt of crazy-ex-girlfriend mythology. Myth confirmed—it was me.

KISSING FROGS

EXPERIMENT

After Jason, I had one really sweet boyfriend in college, but who wants to hear about that? Sorry, Matt, you are not going to make the book and I know you bought it just to find out what I said about you. I'm afraid you were one of the good ones and wrote too many nice songs about me. Sorry!

After Matt, I found myself attracted to other "fixer-upper" boyfriends, strivers, musicians, and brooding artists who were chronically under- or unemployed and needed a cheerleader/problem solver to help them realize their dreams. The role fit me well, I'm ashamed to admit, although it was far from the independent-yet-romantic fantasies of my youth.

One guy was a vegan chef, an aspiring painter, a dreamer, and a drinker. I didn't think his art was that good, but he thought he was a genius. He had so much confidence and ambition, I thought maybe he *was* on the cusp of finding greatness.

While he dithered creatively, I did *everything* for him. I let him live with my roommates and me, lent him money, and attempted to draw out the brilliant in him, but he never got better at painting, or at living. I never even loved him, but I stuck it out for a long time because I was entranced by the romance of being a muse. My friends hated him, and yet I persisted. One plus in his favor: He was a chef. With him, I was a less starving artist for a hot minute.

Once, we were having sex, and I muttered, "Matthew!"

His name was Michael. Matt was the sweet college boyfriend.

The moment I realized what had just happened, I panicked for a second, but then he said, "I love you, too." He thought "Matthew" was "I love you." He was so touched, I couldn't tell him the truth. I'd been on the verge of breaking up with him, too, and now felt stuck. I stayed way longer than I should have.

I was right about his work. He got nowhere with it, but I did as much as possible to help him realize a dream that was doomed to fail. He ended up running off to Indonesia with the money I'd lent him. He left all his shit behind and told his landlord that I would deal with it.

Why was I so committed to the idea that love is hard work? I was a soft touch, just like my father who was always bringing home strays, people which both of my parents were pained to get rid of. The man would lend you his last dollar if you needed it!

I knew what was going on, but I operated with the theory that it was greater to give than to receive love, and acted it out. I took care of a lot of sad cases who didn't love me back the way they should've and learned that there is no end to men who are on the road to figuring themselves out and in need of a funny, artsy woman to listen to them for hours and run errands while they do whatever they need to do instead (read: play video games). Instead of encouraging and helping these guys, I should have been cheerleading for myself.

I call men like this "emotional vampires." It's easy for emotional vampires to find people like Dad and me, and say, "Hey, I can suck the life force from them for a while." When Michael left, I vowed he'd be my last stray dog boyfriend.

HOW TO DUMP A GUY

I was the dumpee as often as the dumper, and my behavior was no better than the jerks who left me. It might sound horrible, but I'd rather get dumped and wallow in misery than do the dumping and be tormented by guilt. I found it nearly impossible to walk away from people even when I knew they were wrong for me. I was always so worried about hurting them, I'd pretend everything was fine and hide my true feelings while slowly distancing myself until—*poof!*—I was gone. I ghosted before that was a thing.

The Wrong Way to Dump a Guy?

What I did, the stealth withdrawal and sudden disappearance, looking back, I see how cowardly it was. I regret not having the guts to share my true feelings with men I'd been close to, even if it was to say, "I don't love you anymore and cringe when you touch me."

Don't get me wrong, I'm not crying in my coffee about it *now*. I didn't scar anyone for life (that I know of). I just wish I'd been brave and given the sweet boys whose hearts I broke the respect of being honest with them.

The Right Way to Dump a Guy?

Sit down with him, look him in the eye, and say, "I don't love you. This isn't going anywhere. I'm sorry but we have to move on." It'll hurt you to say it and for him to hear it, but truth is always less painful than a lie in the long run.

Besides, staying with the wrong guys is such a waste of time. Science tells us that a larger sample size will yield better results. I say, kiss as many frogs as you can.

LOVE IN THE BEER AISLE

San Francisco, in case you didn't know, has several miles of magnificent beach. One night in 1999, I wanted to go down to Ocean Beach to check out a big protest—the offense du jour, the Enron-spurred energy crisis and resulting blackouts. In SF, most protests turn into rave-like parties, especially on the beach, with bonfires, drum circles, singing, and drugs, lots of drugs.

I called my best friend Lisa's house to ask her to go with me, and some guy named Paul answered the phone. I'd seen him around once before. He had crystal-blue eyes and the long, blond dreadlocks of a surfer Rastafarian, which he was. I passed him once standing next to his Harley, smoking a cigarette. I was rocking a happy Goth look in my white leather trench coat and ornate black eye makeup at the time. Our styles did not mesh and, admittedly, we were not each other's type. Opposites really.

Anyway, he said Lisa wasn't around, so we hung up. Then my phone mysteriously rang. Back in the day, you could dial *69 to call the last number that had called you. It was Paul, calling back to say he thought checking out the craziness at the beach sounded cool. I casually suggested he meet me at Safeway in the beer aisle (as one does). Half an hour later, there he was, next to a stack of Budweiser cases, this tanned surfer in a quilted jacket. Funny how details of important moments stay with you—the fluorescent lights from above glowing on his face, the uncomfortable eye lock when we couldn't stop staring, and then nervously looking away.

I was struck by the sight of him, and surprised by my reaction.

I had another guy with me, a friend from work. The two guys looked at each other, confused, while I grabbed a six-pack and brought it to the register. "Are you coming?" I asked Paul.

Lil **Q**   *Why Is Eye Contact So Powerful?*

When you look directly into someone's eyes, it can mean many things. A warning. Proof of trustworthiness, as in, "Look me in the eye and say that." It's how you show empathy, concern, attention, or send a secret message. Prolonged unbroken eye contact for two minutes is almost all you need to fall in love.

According to a study by researchers at Clark University, random strangers of the opposite sex (hey, the study was done in 1989, pre-genderfluid acceptability) were tasked with gazing into each other's eyes for two minutes. Afterward, they reported feelings of love and arousal, despite having only just met their experiment partners. Apparently, eye contact releases phenylethylamine and oxytocin, chemicals that make you feel attraction and bonding. If you're looking for love, definitely keep your eyes open.

Hesitantly, he said, "Okay," and the three of us took our beer down to the beach together. I didn't realize our threesome was super awkward. I assumed both guys just wanted to be my friend, but I was wrong, times two. Paul had sussed out the situation and didn't want anything to do with it. We were at the protest/party for only a few minutes when Paul said, "I'm going to the Kilowatt," a bar in the Mission.

I said, "I want to come! Let's all go together!"

So I made this poor guy from work, who thought we were on a date, drive all three of us down to the Kilowatt. I was focused on Paul completely, and at some point, the other guy disappeared. I guess he realized he wasn't getting anywhere.

So there I was, in a bar across town with a surfer skater guy that I didn't know except that he was a friend of Lisa's boyfriend. None of that mattered. I was chasing a moment, feeding the butterflies in my stomach (mostly Jägermeister) while I tried to seem super cool so Paul would like me.

We ended up talking until the bar closed, and, with no car, decided to walk the miles back to our neighborhood. We arrived at my door at 5:00 a.m., walked and talked out. Both exhausted, I let him sleep over at my house. Nothing naughty happened because I really liked him. I was thunderstruck and playing for keeps.

In the morning, I woke up face-to-face with him, breathing in his breath like I was consuming his presence. The symbolism of the moment was so beautiful that it even overrode morning breath. At some point during our intense twenty hours together, I'd fallen in love. **To find love, look in the beer aisle, or in any unexpected place.**

Another Lil Q **Morning Breath, WTF?**

Morning breath is, basically, the stinky by-product of normal mouth bacteria breaking down errant food particles in your teeth. But when you're in love, according to my personal research, morning breath doesn't smell so bad. Love is blind. It's also anosmic.

Over the next month, we spent every single moment together. For once, I didn't bend over backwards to please a guy, and he didn't ask me to do anything except to stop trying so hard, and to just be myself. Paul didn't bring out my fixer-upper impulse, or the pressure to prove myself to him. He was happy with me in jeans and a T-shirt, and I relaxed into the style, knowing I was sexy no matter what to him. We were both broke artists and didn't have money to go out, so we'd have romantic nights at home, picking a portrait from a photo book or magazine and drawing it side by side on the couch, and then showing each other our interpretations. It was so exciting to be with a man who loved my art and my mind, and pushed me creatively and intellectually.

WHEN IT COMES TO LOVE, NOT WORKING AT IT WORKS BEST

ANALYSIS

Our relationship had to be kept a secret from our friends, though, because Lisa had previously warned me against him. "Don't date that guy," she said. "He's cranky."

Lisa's boyfriend Brett warned Paul against me, too: "Don't date that girl. She's crazy."

So Cranky and Crazy kept it under wraps until we were sure it was something real. It was obvious. We never left each other's side. Paul waited five years to propose. He even had the ring in his pocket one night when I had too many beers and went on a rant about how marriage was just a piece of paper, blah, blah. Poor guy. (I can be such a pain in the ass.) He walked around with that ring for another year before he got the nerve to ask the "I never want to get married" girl for her hand. Funny how quickly I abandoned my stance when he got down on one knee. I cried and said yes immediately.

HOW TO CRASH TEST
A WEDDING

For us, the only important elements of a wedding were having a good time with good people. The planning was thrown together with barely any effort or much thought. The wedding industry sets you up for crazy expectations, which, logically, drives people insane. I wasn't going to get sucked into that mind-set. I also couldn't afford to.

My plan was "go random." To choose a location, I Googled "paradise and pillars." I love pillars for some reason, and figured it was as good a criteria as any, the internet's answer to spinning a globe and putting your finger on a random destination. The first picture that came up was Costa Rica. I had never been there, but booked it blindly anyway. Only the people who really cared would come, and a destination wedding would keep our invite list small. I come from Catholic folks. We have a lot of cousins. (No offense to the family but Grandma Byron was one of seventeen kids! Seriously.)

I walked down the aisle in a white dress and Vans with "Thug Wife" written in crystals on the top. Paul was shocked that I wasn't wearing something red or vampire-like. My dad escorted me in the pimp hat I bought at a zoot suit store in the Mission District. Paul and I wrote our own vows. Our sisters stood beside us as Maid of Honor and Best Woman. My best friend Brittany officiated, and we were married "by the power of Grey Skull." At the reception, both Brett and Lisa took credit for introducing us.

Here we are now, sixteen years later, married, with a kid, a mortgage, everything I thought I didn't want.

In the digital age, most of my friends rely on the mathematical algorithms of dating sites to find love. If Paul and I had plugged in our preferences, we would have been a zero percent match. We like different music, and don't watch the same TV shows. He's a Southern red meat lover, and I'm a California salad eater. But we have a powerful, base-level animal attraction. It sounds weird, but I was drawn to his smell. It was nature, and I didn't have a choice.

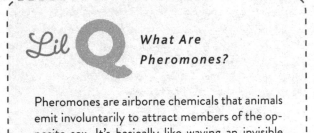

Lil **Q**   *What Are Pheromones?*

Pheromones are airborne chemicals that animals emit involuntarily to attract members of the opposite sex. It's basically like waving an invisible flag that says, "I'm here and ready to do it." In the animal kingdom, animals mark their territory with pheromones. If you've ever seen a cat rub all over a table leg, that's what's going on. Scientific research has proven that pheromones are the reason women's periods in the same household sync up. (Every woman at *MythBusters* was on the same cycle.) As for whether human pheromones cause people to be sexually attracted to one another, conclusive research is pending. According to the empirical research of one couple— Paul and me—pheromones are real and powerful. As soon as I was in his smelling range, I was drawn to him like an animal in the wild.

ARTISTS IN LOVE

When Paul and I started dating, San Francisco was undergoing an artist renaissance in what was known as the Mission Style of painting and sculpture in the '90s and early aughts. One of my favorites was Margaret Kilgallen, a folk artist who was, fortunately for collectors, prolific during her short career.

She married Barry McGee, another artist in the Mission School, and became pregnant. They were both successful artists, in love, and about to bring their dearest collaboration into the world. And then, Kilgallen was diagnosed with breast cancer. She made the heartbreaking decision not to have chemotherapy to protect her baby. She gave birth to Asha, a daughter, and died three weeks later in 2001, at thirty-three, leaving Barry to raise their child on his own.

Paul and I were also two SF artists, falling for each other while this tragic love story unfolded. It was a highly charged emotional time, and it infused our relationship with gratitude. We would not take each other for granted, not for a day.

Even after Paul and I were solidly a couple, we continued to chase each other, and in many ways, we still do. We will always be smitten boyfriend-girlfriend at heart. We keep the romance alive by seducing each other with the tools of our trade. We are artists and we draw. We write. We create. We put our love on wood, on paper, on canvas. We make art about love that engenders love itself.

We painted and crafted postcards out of paper and wood for each other early on, and sent them through the mail to each other, and have continued the tradition. We've sent hundreds of them, and keep them all. We have been passionate letter writers as well, on paper and through email. Paul is a soft-spoken Southern boy who might hesitate to say things, but he can write them! When I closed my Yahoo! email account and went through some of our stuff, I broke out in a sweat and was so glad I'd never been hacked! It was *dirty*! I hope my daughter never writes like that!

Two Christmases ago, Paul wrote me 365 love letters and gave them to me in a box so that I'd start every day of the year reading a message of love. (I know. I'm spoiled rotten by him.) Some were short affirmations, like "I fell in love with your smile," "I love you more than a fat kid loves cake," or "I love you more than Kanye loves Kanye." Some were longer about things we'd done and shared. Some were drawings or Polaroids of him doing something funny. Each letter and photo was unique, and effortful. I couldn't believe the amount of time, thought, and energy it took to create this gift.

And then he did it *again* last Christmas. Another 365 original love letters for me to open every day of the year, and this time he made them for my daughter, too.

The man is a hero among my girlfriend community. He's not so great about traditional holidays. He has to check my Wikipedia page for my birthday and our anniversary and often forgets them both, but who cares? He's an unconventional person and thinker, and that's how I like him. As much as I love my husband's body, I love his mind more . . . until I see him in a wet suit, and then it's back to his body.

This year, he warned me not to expect another box of Christmas letters. "I don't want to be predictable," he said. As if.

For us, love wasn't math. It was chemistry.

Since there was no choice about our being together, I didn't have to worry about whether he liked me a certain way or not, and all of the pressure I usually felt in a relationship was gone. **When I stopped playing a role, I found the movie ideal of love.**

With Paul, I could be a vibrant independent woman who was unafraid to be romantic and in love. Paul was my breakthrough. He didn't want me to be anything other than me. I never felt like I had to be his muse or his savior. He was too cool for that. Once I got rid of the idea of "type" or "boxes," I realized he was the whole package. He was sweet, smart, kind, loyal, and cool—the perfect combination of the bad boy and the smart class clown.

PAINT THE ROOM

CRASH TEST

It sounds very romantic when I tell you about the postcards and Christmas letters, but Paul and I are a married couple and there are periods of time when we hate each other. Like every marriage ever, we fight over money and jealousies, the state of the house and who really does all the cleaning. I *know* it's me (*winking*).

One excellent piece of advice I got when we first got married was, "If you're fighting, paint the room."

There was a time when the walls of our living room were all a different color.

Having a kid is hard on a marriage. Lack of sleep and the stress of responsibility can throw the most perfect couple into a tailspin.

When Stella was three weeks old, Paul went racing on his motorcycle and crashed on the track. He broke his collarbone and couldn't get surgery for three days. After the initial shock and being grateful he was alive, I got resentful. I was still recovering from an exceptionally long, hard labor. I like to describe it as a wolverine punching her way out of me for forty-two hours. Along with postpartum healing, I was breastfeeding around the clock and exhausted. Paul was stir crazy and throwing a fit of Peter Pan rebellion against the newfound weight of being a dad. If he had come home from the race unbroken, maybe I would have let him off the hook with a couple of passive-aggressive personal jabs. But suddenly, I had two helpless creatures to take care of. No, make that five: We also had two puppies and an incontinent dying old dog that shit indoors.

I bit my tongue until his collarbone was tended to. But the moment he was feeling better, I braced myself, and we had the biggest fight of our lives.

And then we painted.

Painting is a job you can do without talking, but you are still working together on a common goal. You have to keep going until it's done, and then, you feel proud of what you've accomplished and can't stay mad. During our rough patches, this house has been very colorful. These days, the walls are finally starting to match. **By setting a goal—not painting a wall, but forgiving and moving on—to achieve together, you will get over anger and resentment faster.**

Our other make-up strategy is to get sweaty and dirty . . . on the bike trail.

One time, Paul and I were riding in my hometown, and we took our regular bikes off road, which you're not supposed to do. While we were on the trail, we saw this group of guys on dirt bikes coming down a hill really fast, and I saw my husband's eyes light up. He said, "We should try that."

We bought cheap mountain bikes, and started riding together. Our adventures were like marital therapy. If we started out angry at each other, annoyed, or any other negative emotion, we'd be clear of it by the end, thanks to the adrenaline, nature, and shared experience.

I started to really like the sport and did it on my own to relieve stress and alleviate anxiety. Paul went on to become a professional mountain biker. We still go on our nature rides together, especially if there is something we need to work out between us. It always gets resolved, often without speaking, while on the trail. **In marriage and biking, ride right through the mayhem.**

TRUE LOVE IS THE SMILE THAT SMILES BACK

CONCLUSION

Just like our plant-growing experiment, when it comes to love, talking—even yelling—helps you grow stronger. Whether it's "I love you, love you, love you," or "I hate your fucking guts right now," as long as the words are honest and from the heart, you are growing

51

together. When it comes to plants and love, take time, pay attention, and pour out your heart, in words, every day, with affection and/or anger. As long as those vocal vibrations are happening, the relationship will grow.

The best part of my romance with Jon the band geek was talking about it with my friends. During my fix-up years with vampire boyfriends, I didn't speak honestly with them about my feelings—nor did they with me—and it left a lot of bad vibes and wasted energy behind. Kissing a lot of frogs was useful trial and error experimentation, but not for finding a lot of love.

And then, I met Paul and we did nothing but talk with genuine fascination and honesty, and I finally felt strong in a relationship.

MONEY

For the last several years, I've been covering the Super Bowl of backyard engineering at a crazy event called Punkin Chunkin. One hundred thousand people come to a cornfield in Delaware and stay up all night to set up their air cannons, centrifugal force machines, trebuchets, and medieval battle-size catapults like something from *Game of Thrones* that could take down a castle wall. They can hurl pumpkins for a country mile, which is the same as a regular mile but in a cornfield. Most of the competitors aren't trained engineers. They're guerrilla engineers, people from all walks of life who are just really, really into it and love both the insane feats of physics and the carnival aspect of the event, the chocolate-covered bacon and pumpkin-shelled beer. Every person is welcoming and happy to be there, and they love talking about their wacky machines. Since I'm a celebrity geek, a lot of them let me pull the lever and send that gourd soaring.

Another annual event I've covered is Large, Dangerous Rocket Ships (LDRS), a competition to see who can launch rockets the highest and farthest. Some people send up classic rockets. Some launch unpredictable objects, like a coffin, which, at apogee, releases a vampire, or a bride that then spits out a baby. Engineering competitions bring out your crazy, and we need more of that creative crazy in our world.

Everyone at Chunk and LDRS is steeped in their passion for weird, and I love them for it. They're just backyard engineers and makers who have a passion for alternative extreme building and have somehow found their tribe of fellow hobbyists. The vast majority of them are not necessarily rich, but spend many thousands of dollars each year on their rigs. Why?

What Is Money?

Los Gatos, California, today, is mostly an entitled enclave where Silicon Valley millionaires (and billionaires) live in mansions. On the main street in town, you can shop for three-hundred-dollar tank tops and watch a parade of brand-new BMWs, Mercedes, and Range Rovers crawl by. Back in the '80s, when I grew up there, it was a free-spirited middle- and upper-middle-class suburb where hippies who made a decent living settled.

My parents' friends looked more like bikers or surfers and they all partied like it was 1999. My parents picked Los Gatos to settle and raise a family because the community was tight. Everyone knew everyone. The schools were great and you didn't have to lock your doors. It was picture-perfect Californian suburbia, and the Byrons were in it for the long haul. Their vision quest was to be forever 95030.

I was in awe of my father. The one way I tell parents how to get their kids to be interested in science is by being interested yourself. I became a maker by watching my dad in the garage. It must be where I first fell in love with the smell of sawdust. He could have been a professional woodworker with the kind of talent he lent to his hobby. His passion for it fascinated me. He made our Christmas toys, built a beautiful rocking horse, made jewelry boxes, curio cases, furniture, and even the most amazing tree house with a wraparound deck. Working with your hands is an art that I watched and absorbed. It was his way to show us love and I carry that love to my daughter.

My mother is beautiful, and smart, a Berkeley grad with degrees in anthropology and French. The daughter of a doctor and a nurse,

First crash test dummy
(little sister Summer)

she was intrigued by anything medical, and she seemed to know everything, as if she were WebMD before there was an internet. She should have been a doctor, but after modeling a bit, she had me, and then my sister. She poured all of her left-brain, detail-oriented mind into making grocery lists with maps of the store with a symbol key at the bottom and color coding for sale items. Our shopping runs were strategic, planned, and efficient. She watched every item rung up and kept the checker in line. There were no impulse buys. My sister and I followed behind her cart, marveling at her surgery-like shopping precision. Her coupon box was an accurate filing system, organized by category and expiration date.

As successful as my parents were as human beings who embraced love, fun, and the joy in doing, our family's finances were not always stable. My father, a real estate broker, did well when the economy was robust, and we lived in the mountains in big, pretty houses. When the market crashed, we moved to the "flatlands," as Dad called it. My sister recently calculated that we moved every two and a half years, but always stayed within the borders of Los Gatos. Our houses never really felt permanent, like we were going to put down roots. But I got used to moving, and each new house was exciting. I would run from room to room checking the place out, and Summer and I took turns choosing bedrooms. Being a manipulative older sister, I always got what I wanted by talking my sister into what her choice should be. Daughter of a Realtor, I guess.

Dad got it in his head that you had to *appear* successful to be successful. So, even though we'd moved to a two-bedroom apartment, and Mom was scrimping and saving to stay on budget, Dad went out and bought a Rolls-Royce Shadow to drive clients around in. Mom didn't talk to him for a very long time after that purchase. He still has that car, and it's become like the lamp in *A Christmas Story*, a hated object for everyone but him. In a way, I appreciated the dreamer in him, but I was with Mom on this one: I still can't stand that stupid car.

QUESTION

WHAT ROLE SHOULD MONEY PLAY IN LIFE AND IN LOVE?

During high school, our family finances ebbed to extremes. We lost our house and didn't have the savings to move to a cheaper one. My parents, sister, our dog Frasier and cat Tiger, and I moved into the guest room of my friend Lara across the street. Not the guesthouse. The guest *room*.

Needless to say, the guest room, with four people and two pets, was cramped. I tried to make my own space by living in a closet-size office on a mattress on the floor. It was a tough time. Very demoralizing for my parents, and embarrassingly devoid of privacy for my sister and me in the prime of our angsty teenage years.

I felt bad for them, for all of us, during our rough patches. My mother's coping strategy was to buckle down. Her scrimping and coupon clipping took on a frightening desperation. To this day the woman keeps Tupperware in her purse to take leftover anything home for future meals.

If I made my own money, the hypothesis went, it would take the pressure off my parents and give my sister cash to buy clothes so she could have as much of a normal teenagehood as possible. It was my duty, really, to protect her. I've always felt like Summer was my responsibility. When she was born, I even told my parents, "This is my baby. *Mine.*" I was three and a half, so she was my living doll.

I took a series of part-time jobs to do it, and got into the romance of being a seventeen-year-old who supported herself, like a girl in an after-school special. I wanted to know that I could do it, and that being okay was within my power to achieve, not only for myself, but for my sister, too.

Eventually, she moved out of the guest room and in with her best friend. The guest room occupants were now me and my parents (and the pets). The space was unbearably claustrophobic for me. I was dealing with the stress of being a teenager thrust into a bad situation with zero privacy, and my parents' anxiety on top of my own. It was too much. I had to get out of there and deal with what was happening to our family on my own.

I told my parents I was staying with friends, packed all my belongings into a single duffel bag, and drove off in my '80 Oldsmobile Delta Royale, a big ol' American car. I wound up sleeping in it that night. It seemed easy, so I kept doing it, going to a friend's house to shower.

There was no conscious plan to live in the Olds for a month. I was just winging it. At night, I parked behind the bagel store where I had

HOW TO RIG A
STUFFED DUMMY

Take Dad's old thrown-away business suit stuffed with newspapers and old clothes. Add hat. For the face, old glow-in-the-dark skeleton mask. Just enough for someone to back off. No one is going to mess with that big guy in the front seat. Wonder if I drove in the carpool lane with him if I would get pulled over. He was pretty convincing from a distance.

a part-time job. I rigged a stuffed dummy to look like a big dude was sitting in the driver's seat, then I laid down on the back seat, covered myself in clothes, and went to sleep. It was scary and lonely. I was angry that my life was so chaotic, but, on the other hand, I wasn't giving in to the bad circumstances I was in because I was doing something about it.

I learned how to barter during this period. At the end of the night I could take all the leftover bagels from the store and trade them to the coffee guy for coffee, to the sandwich guy for sandwiches. Whatever was left, I would try to give to the three homeless people I knew in the park—Joan, Pete, and the Cranky Guy. Pete was somewhat consistently sane. I would sit down next to him on a bench, share my bagels, and he'd tell me stories about Vietnam. Sooner or later he would get agitated and I would leave him to his bottle.

I wasn't sleeping well but I needed to be wide-awake at my jobs and to do my homework, so I started taking caffeine pills. One night while driving, fretting and overwhelmed about money, wired and exhausted at the same time, instead of hitting the break as intended, I hit the gas and crashed into a tree. The Olds was like a tank though, and the tree got dented, not the car. I told my best friend Brittany about it, surprising myself by crying. She said, "Okay, enough. You're moving in with me and my dad."

I had to admit that winging it wasn't working. Living in my car wasn't the great adventure I thought it'd be. I mean, it was funny and weird, and I did prove to myself that I could do it, that I was a person who could deal with her problems. But I also realized that my self-preservation instinct caused my driving accident. I was good at wanting to deal with my problems myself, and terrible at letting myself look vulnerable by asking for much-needed help. Self-reliance is the foundation of my money philosophy—still is. But accepting my friend's offer was an aha moment: I didn't have to shoulder my parents' troubles and my own by myself. If I could lean on the people around me, I could get through anything. I gave in to the fact that I wasn't weaker for asking for help.

It's hard to overcome pride. In our Instagram society, no one likes to show the bad, sad parts of life. We only show our best selves. But, sometimes, you are your strongest self when you don't let pride hide a crushing weakness. #nofilter.

Before too long, things got better for my parents, sort of. They were able to buy a decrepit, historic little house on a lot with no foundation, and crumbling beams, minimal plumbing and electricity. It was like a cartoon crumbling house, where, if you sat on one side, the whole place leaned. Instead of insulation, newspapers were glued to the wall. It was a goofy little house with two bedrooms, one for my parents, one for my sister and me. I chose to live on the long and narrow sunporch off of the kitchen. We moved the fridge to block the door and give me some privacy. At night, I'd listen to the rats chewing on the floorboards. I would just lie there, listening to the gnawing every night. It sounded like they were everywhere. We tried poisoning them, but one would die in the rafters and smell so bad, you couldn't use the kitchen for a month. My sister and I found one right in the middle of my room, gasping for air, and it was huge, a foot-long body and a superlong, hairless tail. It seemed like *The Secret of NIMH*; it was terrifying. To this day, rats wig me out.

My sister and I talked about what it'd be like to have a normal life, and we were envious of our friends in nice houses where they had lived their whole lives. We couldn't imagine a life where you didn't have to sit on the washing machine to stop the house from shaking bits of ceiling down. She fantasized about joining the swim and racquet club her friends invited her to for the day. I reacted to the moves by rejecting the very idea of "home," saying I would be a drifter artist who didn't stay in one place for long. I'd go wherever the muse moved me, and never feel tied down to a person or a place.

Despite the chaos, I graduated high school with good grades and I went to West Valley Junior College and then on to San Francisco State University on loans and scholarships. While I was in college, my dad's business picked up enough to tear down the rat house and build a nice one in its place that I never lived in—but Summer did, and she got to experience stability, which made me happy.

I knew my parents were doing the best they could. I never once felt like we were poor or homeless. I was a white girl in a middle-class neighborhood. I was never hungry or without a warm place to sleep. Moving into my car was an experiment I chose to do, a crash test on my survival skills. I had a safety net of extended family and close friends, but it seemed important at the time to see if I could make it on my own. **Rough times show you what you're made of, and, if you cope, you get to take that invaluable knowledge with you as you move forward.**

I found that not only could I get a job, I liked having one. I became addicted to employment and making my own money. I also learned that I didn't need much and could tolerate a high degree of personal discomfort. My homeless-by-choice month was my rebellion against my father's assertion that "it is just as easy to love a rich man as a poor man." He made it sound like I needed a successful husband to secure my happiness and future. I reacted to his ethos by wanting to be the hero of my own story. I said, "I don't care about money!" and proved it by living with hardly any.

And yet, to this day, my deepest fear is eating cat food in a box under a bridge.

Money for me means security and stability, two things that I didn't often have as a kid. Nowadays, I don't need or want a fancy car or the trappings of success. I just need a permanent home for my family. I'm obsessive about not carrying debt. I'm very much my mother's daughter in being an avid budgeter and keeping a close eye on the bottom line no matter how much money I make or have saved.

MONEY = SECURITY AND STABILITY

HYPOTHESIS

Once you take care of your basic needs—food, shelter, safety—you will be okay. I will never equate wealth with happiness, because I've seen and experienced the world from both sides. I've met fantastically rich people, TV execs and entrepreneurs, and tragically poor people in the slums of India and the favelas of Brazil. The people in pursuit of status and the dollar are the most depressed and unsatisfied of all because they always want more. But people who have good relationships, no matter their incomes or the size of their houses, are happier and healthier.

That is the American dream after all. Do what you love and you'll ultimately be financially rewarded. When I did an ad hoc poll of my many lifelong artist and musician friends, their answers were to burst out laughing. Granted, some of them have become successful and they lead comfortable lives. Some of my friends decided that they were passionate about making a lot of money, and they went into real estate or finance, and have followed their bliss all the way to the bank. The truth is, no matter what your dream might be, you have to work your ass off to be successful in any field, including those with more earning potential than "starving artist," which was always number one on my career wish list.

GET A STRATEGY

EXPERIMENT

The Starving Artist Financial Plan

After college graduation and my wander year, I survived on odd jobs and bar food. My mom had given me a map of every free food happy hour bar and free samples supermarket and liquor store in San Francisco. It was like the key to the city. I have always loved eating hors d'oeuvres for dinner, and was fine sustaining myself on nachos and food with toothpicks in it. My favorite place was an Irish bar that had a full Indian buffet every happy hour (aka San Francisco fusion). Mom created this list way before the internet from word of mouth and newspapers. She truly was a freebies queen.

When I met Paul, we were both struggling to pay rent, but even then, our divergent spending styles were emerging. I was a saver, and Paul was a spender. (Paging Dr. Freud.) Financial woe was, absolutely, the number one source of tension between my parents, and in the spirit of not making the same mistakes they did, I sensed Paul and I had to get on the same page about money ASAP.

If I'm weak in an area, I seek out someone who's strong. Back then, I wasn't well versed in finances, so I sought out an expert to guide us. We didn't have much money to spare, but I hired a financial advisor anyway, and chalked it up as an investment in our future harmony and happiness. The advisor walked us through the basics, like making a budget, paying down credit cards, and what compound interest was. If we could save a little bit each month now, it'd grow exponentially as we got older. I was fascinated by it all and realized I had an aptitude for it. What Paul got out of the session was the confirmation that money does not interest him much. So we agreed that I would be in charge of our finances. Paul trusts me completely and backs me on my decisions.

So when I decided it was time to buy a house, Paul was only too happy to not have to worry about all the details and just let me do my thing. I give him credit for his staying calm. I knew he was apprehensive about my taking on such a huge responsibility, but he followed me into the unknown with trust. I probably would have insisted on that anyway, since I'm a control freak about saving, investing, and having a safety net. I'm convinced that Social Security will be gone by the time we hit sixty-five. I've seen people ruined by outliving their retirement savings and am fearful about cost of living increases and how much it takes to save for Stella's college tuition and our old age. I'm on it, though. I'm planning for our future. If I handed the job over to Paul, I'd worry. Not that he isn't capable—like I've said, I'm a control freak. **Money doesn't have to be the root of marital strife if you agree on a strategy that works for you.**

I'm the CFO of our family. Our arrangement is not conventional. People still make assumptions about who pays the bills. Whenever we negotiate with contractors, for example, they always go to Paul for expenses approval, and he sends them to me. (I love getting mansplained about construction. "No, sir, please explain what drywall is to me." I am just a silly girl with a forklift certification and an entire garage of power tools.) When we go out to dinner, waiters place the bill in front of Paul, and he very pointedly pushes it toward me. My husband is a "fight the power" person and jumps on any opportunity to raise awareness about inequality. That's my feminist strong man.

SMART SPENDING

I was once speaking at a conference that was aimed at inspiring girls about the future. As I was waiting side stage I caught Nely Galán's speech about financial literacy. She emigrated from Cuba with nothing and went on to become the first Latina president of Telemundo. In her book, *Self Made: Becoming Empowered, Self-Reliant, and Rich in Every Way*, she dished out tons of excellent advice for women in business—such as "Don't wait for a prince charming" and "Power is taken, not given"—but her tip that resonated with me is "Buy buildings, not shoes."

I respect that immensely. We have to be smart about spending, which means saving and investing wisely. Buying frivolous things will not get you anywhere except, to quote a Johnny Cash song, "Another day older and deeper in debt."

We live in a strange culture, where looking glamorous "means" you're successful. When I see a super-glamorous young woman who cares more about her hair than her, um, like, *diction*, I see a sucker.

No one cares how fancy your purse is besides you.

Money is power. By spending it on things that don't last or increase in value, you fritter away your power. Before I spend a single dollar, I ask, "Will this pay off?" Every purchase is an investment, so I invest my money on:

- **Good tools.** If you get a crappy, cheap band saw or drill, you'll have to replace it soon, doubling the cost. If you buy quality first, you'll save in the long run.

- **Good boots.** Look, I like a nice shoe, and I have splurged on some designer heels that won't be in fashion forever. But I live in boots with good arch support and steel toes. A well-cared-for pair of Frye boots will last a lifetime.

- **Good accessories.** In my twenties when I could ill afford it, I spent one hundred dollars on a leather belt. I have been wearing it for twenty years, which retroactively makes it an absolute steal.

- **Travel.** Travel is the only expense that makes you richer by experiencing new things, seeing the world, gaining a perspective on your own life by seeing how others live, learning to talk to people from different cultures, and sharing ideas. You will be wealthy with insight, knowledge, and point of view. One day of walking the streets of India, and I promise you will come home and not want to waste a minute of your time on bullshit.

- **Food.** When I stopped surviving on free bar food and made enough to go to restaurants, I was shocked to see how much we spend eating out in a year. So I learned how to cook, and learned how important quality ingredients are. Every purchase is an investment, and there is no bigger investment than your health. Call me Californian, but I know I'm stronger when I eat local, seasonal, and fresh.

- **Time.** When you feel low, invest time in something that will make you feel better. Chances are, it'll be inexpensive, like cooking a meal, singing, or dancing with someone else, going to a movie, taking a bike ride. Put positive energy into the universe, and it will come back to you, with compound interest.

Mythbusted: If You're on a Hit TV Show,
You Must Be Rich!

When I was offered the role as on-air cohost, Paul said, "Holy shit! We're gonna be rich!" I thought we would be, too. How wrong I was.

When I started as the cohost of *MythBusters*, I was paid four hundred dollars a week for working thirteen-hour days, five or six days a week. After the show took off, my salary stayed the same. I was still taking out the trash every day at M5, and glad to do it. I loved the work and couldn't believe I'd lucked into it. I'd signed on for an unpaid internship, so any cash was a bonus.

For the first four years of being on a hit TV show, even though I was slowly making a little more each season, I was still living below the poverty line. I started to think *someone* is getting rich on the show. The reality of reality TV is that hosts are expected to supplement their income with endorsement deals. In the meantime, it took me three paychecks to pay my rent on a one-room studio apartment in a bad neighborhood.

The fabled endorsements and sponsorships did come through eventually, and, much to my amazement, I made the equivalent of a year's salary in two weeks by doing a promotion for a big company. Suddenly, I was making six figures and working even harder, motivated to strike while the iron was hot.

The best part about making more money than I'd ever thought possible was not feeling the constant, nagging stress about being able to take care of myself and my family. I was also relieved to be in the position to help my friends in need, and I had a lot of them.

I was *not* so thrilled that every distant relative I'd met once came out of the woodwork and asked me for money. People I wasn't related to and didn't know wrote to me and said, "We met when you were a little kid. I've hit hard times. Can you send me five thousand dollars?" Complete strangers emailed me, too. A typical message was something like this: "Hello, Kari. I'm a maker like you and I need money to build my robot. Can you help?" People I went to high school with friended me on Facebook solely to ask me to invest in their ventures or donate to their school auction.

I established a policy early on not to reply to strangers. But I felt obliged to write back to family and tried to explain that I wasn't in the position to send large checks at will. I was very careful about the wording so I didn't offend or alienate them. Sometimes, they wrote back, "This is family. You have to send money." I was put in a sticky and uncomfortable situation. It would have been sanctimonious of me to say, "Hey, when I was broke, I worked three jobs or sold stuff. I got my hustle on." I didn't do that, but the idea of giving people advice that might actually be of value became a calling of mine. I couldn't give money to everyone, but I had something of value to share: my time. I would offer to, essentially, "teach a man to fish," and occasionally, people took me up on that.

ANALYSIS

BECOMING FINANCIALLY SECURE IS A JOURNEY OF SELF-EMPOWERMENT; TO FEEL THE FORCE OF IT, YOU HAVE TO LEARN TO MAKE IT, SAVE IT, AND SPEND IT WISELY ON YOURSELF AND OTHERS

The Standoff

Until I worked in TV, I didn't hypothesize that my salary was a measurement of my worth. But that changed when I learned that Grant and Tory were paid more than I was for the same job. I don't remember exactly how it came out, just in casual conversation. I do remember feeling shocked and a bit betrayed by the production team for shortchanging me.

I brought it up in my next contract negotiation. The woman I was dealing with said, "They're paid more because they have higher degrees than you."

"Actually, they don't," I replied. Tory and I had the same degree from the same college. Grant had a bachelor of science degree; he was an engineer, yet he didn't have a master's in it.

She had a few other arguments, but I pointed out we all had the exact same job title and in fact I had seniority. So why was I paid far less? I was so frustrated by this conversation, and how she tried to shoot down my logic, that I almost cried on the phone. I held it together because I didn't want to throw gas on the fire or show any weakness. The worst part about the entire conversation was that she was a woman. Why wasn't she championing me or at least offended by the absurdity of the pay gap?

After many conversations about the situation with several executives, the production team finally raised my salary to their level . . . and then they gave the guys a raise over that! Of course, the guys told me right away and were just as appalled by the inequality as I was. Tory and Grant are glorious feminists and were totally on my side during the entire dispute.

Enraged, I hired an agent and went to war. Grant, Tory, and I joined forces and negotiated a "league of all nations" contract that ensured we were all paid the same amount.

This story isn't atypical of other women in TV, or anywhere. Gillian Anderson had to wage war on Fox TV to get the same salary as David Duchovny on the original *The X-Files*. And then, when they revived the show ten years later, the execs pulled the same crap again! She threatened to walk, and I applaud her for standing her ground. She got parity. **Equal pay for equal work is far from a reality, despite feminists fighting for it for a hundred years. And it's never going to be, unless we persist.**

Inequality at *MythBusters*, and other jobs since, has turned me into a wage warrior. If I am doing exactly the same job as a male cohost, working the same hours, why should his salary exceed mine? I'm not asking for more. I'm asking for fair.

I felt like it was my duty to start a new standard. My agent adds a clause to all my contracts that says if my cohost has the same job, same title, same billing, he cannot make more money. I've come close to walking away from the table over this clause. Goes without saying, men don't have to risk jobs for equality. I do it, in part, to give back. Women that come after me might have a slightly easier battle if I have already taken a little of the fight out of the decision makers and disrupted precedent.

It all began when Grant, Tory, and I compared salaries. If they never told me what they made, I wouldn't have gone to war, hired an agent, and joined forces over it. In talking with other women in my field during this time, I learned that not only were they making less than the male colleagues, in many cases, they were often more qualified and experienced and performed at a higher level. From production assistants to producers, the story was all too familiar.

Outraged, I started working with other women, giving advice and counsel about their salaries. I'm giving a hand to women on the way up, and helping those trying to break in, whether it's offering negotiating tips or making introductions. I carved out some pathways, but it's not necessarily easier to be a woman in science entertainment. We have to support and inform each other.

The first step is to talk openly about money.

The bosses perpetuate the idea that salaries should be kept secret because they have ulterior motives. If you don't know what your colleagues make, you don't know to ask for the same. We've all been indoctrinated to think it's rude to talk about what you earn, but isn't it *far more* offensive when women are routinely paid a fraction of their male counterparts' salaries?

From the get-go, men are offered higher salaries, and then they push for more. A pushy man is a smart negotiator.

A pushy woman is a pain in the ass.

NEGOTIATING
WHILE FEMALE

Some tips about how to get what you deserve while interviewing:

- **Act like you're hiring them.** You get to judge, too. Is this the right place for you? Do you really want it? Sussing out whether you like them takes the pressure off trying to make them like you. A good job is about the right fit. Ask questions about the office culture and how they do things to gather necessary intel and to show interest and respect. Bosses and HR people can smell a suck-up a mile away. It's far better to come off as curious and confident by asking what you can do for each other.

- **Smile.** I can't stand it either when a random man on the street tells me to smile. But on a job interview or negotiation, smiling is a power play. I smile through pain, through anxiety, through depression. It's a way to trick your brain into thinking you're okay and in control. Smiling is the least you can do for the greatest impact.

- **It never hurts to ask.** When you are being offered a job, ALWAYS let them make the first offer, and then don't accept it. Ask for an additional 10 to 30 percent right off the bat. They might turn you down, but if you don't ask, you'll never know.

- **Get sweeteners.** Sometimes, salary comes out of one budget, and other perks come out of another. If they say they can't pay you any more in salary, go for sweeteners, like an expense account, more vacation days, a year-end bonus, a guaranteed raise in six months. I always try to pile on, asking for hair and makeup and a wardrobe reimbursement, to travel in business or first class, for a percentage of back-end profits, the bulk of my fee up front. Once, to sweeten the deal of my appearing on a reunion episode, producers gave me five pounds of the black powder that I use to make explosive art. A typical order for *MythBusters*, but not at your local hardware store.

- **Have a "fuck off price."** Would that we were all in the position that my famous friend who shall not be named is often in, when he's asked to do a job he's not that into, and quotes a ridiculously high amount, what he calls his "fuck off price." If they are willing to pay it, it's worth doing and he'll take the job. Otherwise, he gets out of it without having to negotiate or feel awkward about saying no.

- **Be prepared to walk away.** Women have been trained to compromise and be people-pleasers. Don't give in to the impulse to make everyone happy during a negotiation. Know your worth and demand it. If you don't get a respectable offer, throw down an ultimatum, and mean it. I've been told, "Take it or leave it," on certain jobs, and had to say, "Leave it." I walked away with heart palpitations and thought I might throw up, but I walked. The next day, they called and said, "Okay. We'll meet your number. We start shooting tomorrow." Even if it doesn't go your way, you walk away with pride, which money can't buy. You lost a job, but not your standards.

I would love to go Norma Rae on this issue and stand on a table in a sweaty T-shirt, holding up a sign that says UNITE! We have to band together and gather data about what people make and what numbers women should push for. If you have information, you can go into any negotiation prepared. Just this week, I was talking to a woman in my field about freelance fees. I told her what I make on a gig, and gave her the contact info for my agent and suggested she ask him what she should be getting. I've been known to sleuth what kind of money is being offered to my male counterparts or get my agent to listen to the whispers. Then, if they make me an offer, I'm prepared to counteroffer with a fair amount. It's underhanded, but this is war. We have to do whatever we can for fairness.

I think one of the reasons women don't push for what they are worth is because they don't feel they deserve it. Well, you are worth it, and you should always ask for more. Stop being shy or polite. Study after study shows women are less likely to negotiate for more and feel like they'll be viewed unfavorably if they do. Girls are raised to be polite and accommodating, while boys are taught to be self-promoting and tough. Time to stop reinforcing gender roles especially when it comes to money. Don't think of money as a dirty subject. It shouldn't be a measure of worth, but it is a measure of how much you're respected, how far you've come, and how hard you're working. If no one ever talks about inequality, it'll never change. So talk! Be a pushy woman who looks a man in the eye and says, "I'm worth more." That squeaky wheel is gonna get greased.

MONEY CAN'T BUY HAPPINESS

CONCLUSION

I've been freelancing for several years now, and find that when it comes to money, I'm exactly like the Punkin Chunkers and Large,

MONEY CAN'T BUY

HAPPINESS

(but it can buy burritos, and I'm always hungry!!)

Dangerous Rocket Ship builders: My version of bliss is when I'm stimulated and challenged, and price, expense, and income have (almost) nothing to do with it. I'd rather make less (but always equal!) to work with quality people doing quality things. I might not make a ton of money that way, but I will be enriched by positive experiences.

I have taken jobs I hated for the paycheck, and have regretted it each time. Life is too short to work with awful people on a subpar product. Of course, when I take a job that fits the bliss bill, I will push to get as much money as I can and as much control over the content.

My role model for this is the mad, creative genius Steve Wozniak, inventor, engineer, cofounder of Apple, and a friend. When he was inventing the Macintosh computer in his garage, he wasn't thinking about being rich, he was just doing what he loved. Woz is all about what makes him happy, the tech and gadgets and inspiration. He's the player and made his life the game. Granted, he is richer than God, and can do whatever the fuck he wants. But that's my point. He does!

For me, money has always meant security. Once I could pay my bills without worrying, was free from debt, and had my forever home, I was happy. All the rest is just icing, sweet but empty calories. **Save as much as you can and invest in quality to keep your stomach and heart full.**

CHAPTER FOUR

FRIENDSHIP

n *MythBusters*, we did an episode about employing science to get past several security systems. Episodes where I could learn secret agent skills were my favorites. In one episode, we attempted to get past a room of crisscrossing infrared lasers, like in every cliché heist flick. We tried a few approaches, like blowing face powder on the beams to expose and limbo through them (of course, they had me do that one, maybe just because I am the most flexible of the crew and not because I am the only one with a compact), using a mirror to bounce away the beam, and using another laser to turn off the power source.

Another favorite was trying to fool an infrared camera into not registering the body's heat signature. We descended through a ceiling panel and used Grant's sophisticated articulated robot arm with a clamp attached to a small pane of glass to, in theory, block the camera and create a continuous temperature shield, allowing a person to walk into the room undetected.

The problem was, the robotic arm, with its elbows and internal pulley system for maneuvering at an odd angle, just didn't bend the right way to get that pane of glass in front of the sensor. Grant wanted to engineer adding another joint to the arm, and I said, "Just give me, like, a broom or something."

Hanging out of the ceiling panel, I took a broomstick and hung a piece of glass on the end and placed it on the camera.

Guess what? After all the complicated and convoluted attempts to disarm the camera, the most simple and basic idea was the one that worked. We coined it the "poke it with a stick" method.

From that day on, if our prototypes for a rig were getting more and more complicated, someone would invariably say, "Or we could just poke it with a stick."

Friendlessness

When my class moved from elementary school to junior high in seventh grade, it was a precipitous leap from the cubbies to lockers. The junior high in Los Gatos was fed by three elementary schools, and when they all joined up, the whole pecking order of popularity had to be reshuffled.

I wasn't a total loser in sixth grade or anything. In our small elementary school, most everyone was friends with each other. That all changed in junior high. If you didn't notice the shift, it's because you were popular.

During the transition, my standing in the social hierarchy fell off a cliff. In the new world order of junior high, my old friends were the ruling class, and they decided that I was too odd to walk among them. The distancing was gradual—running away from me in gym class, mocking my comments, some sneering—and then all of a sudden. A few months into the new school year, the Queen Bees did their "spring cleaning" (it was November, but whatevs). Behind my back, they said I was no longer friends with them, and that no one else should be friends with me either because I was "weird."

In hindsight, I can almost see their point. I was, am, weird, and proud of it. I just did not fit in. I sucked at sports and wasn't rich or traditionally pretty. On the contrary, I was uncoordinated, skinny, and full of self-doubt about pretty much everything. Plus, my MTV-inspired style was outré, and I loved to draw pictures—a hobby deemed unacceptable by the Regina Georges of my school (and probably yours).

Like a million lonely geek girls before me (and after), I set out to win over the popular kids with cake. For my thirteenth birthday, I invited all the girls who'd dumped me to a party at my new house.

Nobody showed up.

My family and I set up a big table, with cake and candles. We rented videos from the Warehouse, including Justine Bateman's *Satisfaction*. Not a single girl came to see it. I was crushed and mortified . . . and then, the doorbell rang! This one girl who'd been sick for a week and didn't know I'd been spring-cleaned appeared. She seemed surprised to be the only one and excused herself to the bathroom. She took so long in there, I went to check on her. She'd sneaked into my bedroom to use the phone to call the other girls. I heard her say, "I'm the only person here! You have to come to this party!" They weren't interested, apparently.

My parents were devastated for me, too, and said, "Let's get out of here and go ice skating." So we took this one girl to the rink and tried to manufacture some fun. It was super awkward the whole time. I felt so much worse for my parents than I did for myself. I didn't want to disappoint them with my social ineptitude. After all, they have always been the life of the party with more friends than I could keep track of. How did they end up with me?

QUESTION

HOW DO YOU MAKE THE BEST OF FRIENDS?

Girls are cannibals. It wasn't my fault that they rejected me. I wasn't inherently repellant. It's that the mean girls enjoyed hating me. As the mother of an eight-year-old, it's crazy watching this stuff start at such an early age. Boys fight it out. They bond over giving each other a hard time, but their trash talk isn't real. Girls tear each other down viciously, and it can continue on into adulthood.

The party was it for me. My status as a pariah was sealed. I knocked my head into the brick wall of popularity for a while longer, to no avail, before I realized that these girls would never be friends with me again. Those ships had sailed. I wasn't sure why I was interested in being their friend. I guess I just didn't want to give up a fight. Maybe I was fated to roam the earth alone like Caine (obligatory nerdy *Kung Fu* reference). Or not? I could just crash test my way into generating new friends.

CRASH TEST

THE STAKEOUT

I devised a plan. I focused on two neighbor girls—Yurah Kang and Lara Kemper. They were best friends, and seemed sweet. I would watch them leave their houses and meet each other in the driveway, greet each other with big smiles, and then start walking down the street toward school. I mean, come on. We lived on the same street. We were meant to be friends.

HYPOTHESIS

IF YOU WANT TO MAKE FRIENDS, YOU'LL HAVE TO PLOT AND STRATEGIZE TO GET THEM

I would get ready for school super early as not to miss their departure, wait at the window for them to meet, and calculate my approach. Then I would run outside at the appropriate time, and walk on the opposite side of the street with my head down, and hope they would notice me and invite me to walk with them, which, mercifully, they did. *What a coincidence that I was walking to school at the exact same time! Every day!*

Like I said, painfully shy.

I walked along and listened to them talk, not saying much, just tagging along, until one day, I was comfortable enough to join the conversation. Soon enough, Yurah and Lara adopted me into their clan of smart girls. (BTW: I'm still close with both of them. Lara is a biologist, and Yurah works in science tech. I couldn't have been luckier to plot my friendship with them.)

Freshman year, I plotted again. There was a new girl at school who'd just moved to Los Gatos. She grew up in New Jersey. After talking to her for a minute—she was cool, funny, smart, and had an edgy East Coast style that was intoxicatingly exotic—I decided she was going to be my new best friend. I made a real effort to welcome the new girl and include her in my friendship circle, counteracting the exclusion of the mean girls. Before long, we were best friends—and remain so today. Twenty years after I first stalked her, we still meet for coffee every Friday.

In a way, I was like an Agent of S.H.I.E.L.D., assembling a ragtag team of talented, funny Super Friends (am I mixing Marvel and DC? Sorry, fangirls!) to fight evil together. But my posse was snarky, caring, and supportive, and not afraid to be smart and act stupid. I aspired to hang out with anyone who was smarter, more talented, better read, more musical, and snarkier than me, because, even then, I knew that proximity to great people would inspire me to aim higher and be a better person. **You define and are defined by the people you choose to surround yourself with.**

The "One Woman" Problem

I've seen it happen. There's one slot for a woman, and instead of saying, "Why can't there be two slots or a hundred?" we vie against each other. We should band together to dismantle the system that dictates the "one woman" policies.

79

FUN WITH MAGNETS

We used magnets on the show to test myths like whether an electromagnetic James Bond watch would deflect bullets (nope), or if you could use magnetic clamps to soundlessly scale an air-conditioning duct (nope again). Jamie had these insanely powerful magnets, which struck fear in all our hearts. If he walked in the room carrying one, in my mind anything metal would go flying across the room, like a screwdriver or a blade. But Jamie was more diligent about the safety of magnets than even the table saw. You knew a table saw was dangerous. Jamie knew the magnets could also do damage.

The beauty of magnets is their ability to attract and repel. If you line up the positive end of one magnet to the negative end of another, they will stick together. If the magnets were powerful enough, you couldn't pull them apart with all of your strength. Or, if you tried to line up the positive end of one with the positive end of the other, you couldn't push them together. Magnets are used for the power of their attraction and repulsion in turbines to make them spin, creating electricity. They can take off your finger, or create electricity to light up the world.

Does the principle of magnets—opposites attract—apply to friendships? Why is one person drawn to another as if she were pulled by invisible forces? For that matter, why do we instantly take a dislike to someone? A cool thing about some magnets, though, is that once they're heated, they completely lose their ability to attract or repel. So that person you instantly hate? Or the person you are instinctually drawn to, for good or ill? If you get to know them a little better and work up a degree of human warmth toward them, you can judge them without the influence and control of unseen forces.

After the success of *MythBusters*, a few shows popped up that were similar to ours, and there was always the one token girl in the mix. During Q&A sessions on many college lectures, students often asked if I thought the "one girl" on the copycat shows were trying to be me, and whether I was offended.

I was so appalled by the question, I had to pause for a minute and wonder, *Why are you trying to get me to insult another woman and minimize her hard work?* I would reply, "There is room for all of us. I want to see more women doing science and travel on TV."

Having to explain why I supported the women who came up behind me always put a bad taste in my mouth—sort of like choking down my thirteenth birthday cake in the days after that disaster of a party.

Let's be more inclusive, and kind to each other. Let's dismantle the institution of Mean Girls!

Let's show up and make more room for all of us.

Let's take up space.

LET FRIENDSHIP FIND YOU

EXPERIMENT

The Grand Adventure

When I was just out of college, my friend Dawn and I decided to travel the world. We met a travel agent and booked a year's worth of plane tickets. I didn't have much money. I used everything I'd saved or hustled since age fifteen. I sold clothes, hocked anything of value, picked up pennies, checked the change slots in phone booths and vending machines, and even went to one of those "Cash 4 Gold!"

places with all the forgotten jewelry I'd collected from the lost-and-found bins at my old jobs.

I worked my butt off until the day I packed my enormous backpack, hugged my family, and jumped on a plane headed west over the Pacific Ocean. First stop Rarotonga, then Fiji, New Zealand, Australia, Indonesia, through Malaysia to Thailand, Japan, Nepal, India, Israel, Egypt, France, Italy, Spain, Czech Republic, back to Egypt, Ireland, England, and home. I made my way around the world, focusing on Southeast Asia and countries where my money would go a long way.

By the time I got to Indonesia, I'd been on the road for a few months and had become a strange hybrid of punk raver girl (the persona when I left) and granola hippie (who I was becoming). I wore my spiked dog collar and black cowboy hat with yarn bracelets with bells around my ankles.

Dawn and I had so much fun discovering the world together, but parted ways (very amicably) after a few months to find our own adventures. We met up whenever we needed the safety of a companion, but we had become road warrior travelers. I found that I loved traveling alone. I don't mind being alone. Sometimes, I feel lonely around people. I find small talk excruciating. Social scenes and parties with acquaintances are like rusty spoons digging out my eyeballs. It's almost impossible to stand there and listen to people talk endlessly about their schedules and boring routines. I sometimes find myself just nodding and smiling and zoning out completely.

But put me together with a stranger in a foreign city and I am enthralled. When you know you'll never see someone again, you're more likely to fall into a deep conversation, open up, and be honest. A barefoot boy in Budapest and I poured our souls out to each other as we walked the city and castles. We never spoke again, but he may have a bigger piece of me than some in my social circles now.

When I was staying at the Gayatri House in Ubud, Bali, I came back to my room one evening to hear this girl complaining loudly and annoyingly from the room above. I couldn't imagine what could be so wrong with a room that cost a dollar fifty, two dollars if you wanted a bathroom. Later that night, we ended up at the same dinner with a group of travelers.

We sized each other up. Lisa was the kind of girl you picture as the perfect girl from a cheesy teen movie, all shiny mermaid hair and glowing teeth. Turned out, she lived in San Francisco. We left California within a day of each other. She was also spending a year traveling the world. Having so much in common made me instantly wary of her. I'd left SF for a reason, but I couldn't get away from it, or this girl. She kept popping up everywhere I went, like a haunting. I was walking through Kathmandu, and I ran into her on the street. We kept meeting all the same people. I was crashing in Jerusalem with my friend Eli, and guess who he had to go pick up from the airport? Lisa! On her arrival, Eli said to her, "Funny, there is another girl from Bali staying at my house."

She said, "Oh God, it isn't that annoying girl with the dog collar, is it? I swear she is following me."

When I look back on this moment now, I wonder why it didn't occur to me that Lisa was 1) a woman, 2) from the same city as me, 3) doing the same thing as me, 4) in the same order as me. Basically, we were two people who should have immediately sought each other out as friends. I'd had this idea that friendship was something I had to figure out, something I had to select for myself, strategize, and create. It never occurred to me that friendship might find me for a change.

We reluctantly spent the day together walking the old city and fell in friendship. We even got our belly buttons pierced in Jerusalem together. We took trains and buses throughout Israel and made our way down to the shore of the Red Sea in Egypt. By then we had our own language and a cadence to speaking. After we parted and said good-bye, the universe threw us together *again*. We were star-crossed strangers, friends of fate . . . if I believed in that nonsense. I was traveling with my sister in Italy, and I ran into Lisa with *her* sister at a café, and later on the streets of Florence and Paris. We decided that something bad might happen if we didn't stay in touch. Once we grew tired of blowing through our savings in European destinations, we decided to go back to Egypt together. We sailed the Nile in a felucca and shopped the markets of Cairo; a lifelong friendship was now sealed. Side note: I learned that she freaked out about being eaten alive by an infestation of bedbugs in Ubud—my bad for being judgy. I guess there can be something to complain about in a dollar-fifty-a-night room. **Sometimes, fate has to hit you over the head with a sledgehammer. (Literary fate, not real fate, 'cause that is silly.)**

I was a heavy drinker when I met Lisa, but when I fell in friendship love with her, the thirsty void was filled with her and I drank less. A writer and journaler, Lisa taught me to embrace my sadness or loneliness and turn it into words. Emotions, all of them, even the dark ones, were material I could mine for my own stabs at poetry (so bad, really, truly hideous stuff). I had always done it with drawings but never realized how a journal could be both a work of art and a therapy session. We would just sit on the bank of the Seine in Paris and she'd read from her journals about things we'd done. I was inspired by Lisa to keep a journal of my own and came home with a bunch of them. My journals were like art books, with drawings and notes that were a record of my travels—but not only the best moments. It wasn't enough to just see a place, we had to have an experience there that was worthy of inclusion. We pushed ourselves to take risks so that I would have something to document. In the days before selfies, texts, and video recording, we had pen and paper. The words "journey" and "journal" became one and the same to me.

It was all right there on the page: Lisa and I were on a journey together, into ourselves, each other, and the world. We once followed a boy in Cairo to the desert behind the Pyramids and helped smuggle liquor across a border just because he was standing under a huge sign that said TRUST (Egyptians love exotic English words like we love to tattoo nonsensical kanji tramp stamps to look woke). It just made a good story.

84

Lisa and Kari
in Egypt.

WHAT FRIENDS
ARE FOR

Being my friend is rarely boring. You might be a victim of my odd admiration and practical jokes.

For example, my buddy Eric is squeamish about lady stuff, so I sent him a picture of the bucket of placenta from the birth of my baby. I labeled the email "Stella's first house." Knowing that he'd be horrified when he opened an innocent attachment of gory lady stuff was hilarious to me. Eric? Not so much.

When I found out my friend Michael has a fear of clowns, well, that was too easy. I started sending to his house unmarked packages containing my childhood porcelain clown collection. It took him forever to figure out it was me. I delighted in his amateur sleuthing and would say, "I think you're getting close to solving this one, Sherlock!"

I once sent my friend Celeste a box of doll parts that I found in a creepy fire pit. Clearly, she's as weird as me since she baked them in a loaf of bread and sent them back. She is a baker, so it isn't that weird?

So, just a warning: If you befriend me . . . watch the mail.

This is the same Lisa, girlfriend of Brett, who was buddies with Paul. If I weren't trying to find Lisa that night, I wouldn't have met and fallen in love with my husband. I wouldn't have our daughter, Stella, the greatest joy of my life. **You never know who will become important to your life.**

It might be for a single encounter, like that Hungarian boy who lifted me up on a bad day and taught me the romance of a love affair that lasts only twelve hours. It might be forever, like Lisa, who remains a huge part of my life. Allow yourself to figure out what temporary or permanent friends mean for you, and what you are supposed to do for them, and no human connection will be wasted. Everyone is an experience to learn from.

Mythission Impossible: Make Work Friends

Until I started working at M5 Industries, I didn't understand how amazing it is when your work friends are your actual friends. When I found myself at the right place, at the right time of my life, I was compelled to bond with the people there. It wouldn't be like making friends with Brittany or Lisa. Things are a little more challenging when you're the only woman in a shop full of men.

Or so I thought.

Women hear a lot about "leaning in" at work these days. I'm a big Sheryl Sandberg fan, but she didn't write that damn book until long after I'd sorted shit out for myself. When I look back on my time at M5 navigating relationships, I know that **I didn't need to "lean in," because I fit in.** In a room full of shop guys telling dirty jokes, working hard but always down for a prank, I was basically right at home. Literally. It was my natural personality, picked up from my dirty-joke-telling dad, a man who never shied away from a crass punch line, even with his daughters.

I could tell the grossest jokes and swear like a sailor. The shop guys were so surprised by my foul mouth, they were both disarmed and charmed. I appreciated their laughter and attention, and they appreciated mine. It was like cracking the guy code: sarcasm + humor + sexual innuendo = friendships at work. Double entendres rolled off my tongue. While sure, this version of myself wasn't the same one I employed when, say, hanging out with my grandma, it was nice to connect with the people I worked with. And I had a willingness to meet my male colleagues in the middle—to connect with them in a way that put them at ease, and they reciprocated with a respect that put me at ease.

My friendships on *MythBusters* actually reminded me of my high school crew, a ragtag bunch of smarty-pants who inspired me. On science TV, of course, you're going to meet a lot of intelligent, creative people. Jamie has a degree in Russian. Adam is one of the most well-read self-taught people I have known. Grant is an intuitive electrical engineer. Tory has a tremendous wealth of knowledge about all kinds of building and making. As friends, we learned from and taught each other, and each one of them made me a better builder and thinker. Work friendships run deep because you spend so much time together, often more time than with your own family.

In lucky cases, your work friends feel like a family.

The build team—Tory, Grant, and I—were particularly close, and that togetherness created a kind of synergy that allowed the sum total of our individual parts to be greater as a whole, like Mulder and Scully or Woz and Jobs. In certain combinations of personality and talent, something amazing happens. It is something that can't be cast or created, replicated or forced. It's a kind of magic, and we had that on *MythBusters*. A cameraman once gave us the nicknames Practical, Technical, and Logical because, between the three of us and our totally different ways of thinking, we could solve any problem. For that reason, Tory (Practical), Grant (Technical), and I (Logical) were thrilled to work together again on *White Rabbit Project* on Netflix. The format wasn't exactly right for us, and the show lasted only one season, but, man, did we have fun getting the band back together. I hope we work together again, but in the meantime, we remain lifelong friends.

LINDA

Over time, at *MythBusters*, there were more women around the shop and on the production team, and I was glad. One of my favorites was Linda Wolkovitch, a pure genius. She was a researcher and a producer, and her job was, basically, to figure stuff out. She knew every detail of every myth we did. If we needed an expert on medieval torture devices, Linda would find them. If we needed a bomb range or a 747, she could find it. Without her, *MythBusters* wouldn't have had such incredible access. She'd unearth experts who were studying exactly what we were up to and get them to help shape our story. Once, an energetic materials expert she found published a technical research paper on an experiment we did with him, and gave us coauthorship.

For the Pop Rocks and soda myth episode, we needed a cow stomach with the esophageal valve still attached, and we were having trouble finding one. Linda said, "Give me twenty minutes." Half an hour later, cow stomachs with valves appeared. If it weren't for Linda, the true unsung hero of *MythBusters*, I have no idea where we would have gone for two tons of coffee creamer. Even though we don't work together anymore, I still see her when she and other women of *MythBusters*—Linda, Lauren, Yvette, Jax, Jaime, and Francesca—congregate to drink wine, eat organic cheese, talk about the old days, laugh about the crazy things we did, and occasionally gossip about the "where are they now" refugees of *MythBusters* days, while MacGirlvering broken eyeglasses at dinner with a pen shaft and duct tape. Old habits die hard.

People often ask if I was involved in any sexy shenanigans behind the scenes on *MythBusters*. Hate to disappoint, but no. I was a woman among men, but they stayed very firmly in the friend zone. (Not to say others on the show didn't engage in salacious relationships . . . but this isn't that kind of book and it's not my story to tell . . . yet. Maybe someday, *wink*.)

Regardless of my experience, plenty of inappropriate behavior does indeed happen in male-dominated industries by coworkers who should know better—or do know better and hit on you anyway. I have had my share of punishments from bosses that I wouldn't date. Almost every woman I know has some degree of a #metoo story. Some can shake their heads and ignore it; some get to the point that they need to file lawsuits. Hopefully, with more women speaking out, this kind of behavior will become a relic of the past.

What I have found, and what we are seeing now, is the power of women who come together. We are a force when we support each other. And though I adore my many male friends on set, I always feel a special camaraderie with the women I work with.

ANALYSIS

WHEN IT COMES TO MAKING FRIENDS, KEEP IT SIMPLE

How Many Friends Do You Really Need?

According to science, if you have three close friends you choose (as opposed to family) who will be there for you no matter what, you are doing fine.

I have more than three best, best friends, but not too many more.

My social circles are rounded out with good friends and acquaintances to share experiences and make more fun memories. I find different people to tap into different sides of my personality. Having a number of friends with narrow interests won't expand your mind, but having diverse friends with a wide range of interests will. To develop all parts of your brain, maintain a handful of friend groups. I've got my:

> Workout friends for bike riding or yoga class.
>
> Maker friends for crafting and cocktails.
>
> Adventure friends for rock climbing and shooting.
>
> Fan friends who I've met promoting *MythBusters* and stayed in touch with.
>
> Social media friends for DMing on Facebook, Twitter, and Instagram.
>
> Text buddies who watch *Game of Thrones*, *The Walking Dead*, and *American Gods* with me from our respective couches. Doesn't everyone watch a new *X-Men* movie while texting with her nerd BFF (talking to you, Eric!)?

My daughter is my best friend of all time. Our connection touches every part of my brain and makes my entire life profoundly more joyful. But she is not *really* my friend. Friends don't tell friends to stop biting their nails and to go to bed (for the most part). I'm her mom now, but I'm laying the groundwork for our future adult friendship by having fun and working on projects together. I'm sure she'll go through a period of hating me or at least rolling her eyes at me, so she becomes independent, and then, when she comes back, we'll be besties. She is already the coolest person I know.

Poke It with a Stick

When I look around at the wonderful, true friendships that have lasted and lasted in my life, I wonder if all of them didn't emerge from the simplest of circumstances: shared experience, like-mindedness, and respect. Even my first friends—the neighbor girls—who let me walk with them to the bus. While it seemed at the time like I schemed them into being my friends, perhaps they let me join their squad because they too felt high school loneliness, and sensed a kindred spirit.

My work relationships were no different. Despite age, gender, and background, we'd all found ourselves at M5 or *MythBusters* studios, working similar jobs we loved. That like-mindedness made for easy and natural friendship.

The simplest method might be the best one, in busting myths, and forming friendships. When things get more and more complicated, you're not getting closer to an answer or a human connection. You're getting farther away from it. Occam's razor, the scientific principle about problem-solving that says the simplest theory is probably right—it was true when I used the "poke it with a stick" method on the infrared camera, and it's true about friendships as well.

Every time I tried to use complicated measures to make friends, they didn't materialize. But when I applied simple, logical methods like aligning myself with people who cared about what I did, who enjoyed art and adventure, we bonded so easily, it was like we were meant to be friends.

The people right in front of you, the ones you have things in common with, who are at the next desk, in your favorite classes, and go to the same comedy clubs, should be your friends. You might go looking high and low for the people who fill your heart with joy and belly with laughter. But chances are, they're right in front of you. Your friends might not be what you expect, but if someone appears in your life and you enjoy each other, he or she is probably exactly what you need.

One more simple friend formula: The work you put into friendships is matched exactly by what you get back. If you stop working on maintenance, friendships will fade. If you keep in touch, they will sustain. When you have an all-consuming job, and kids, and a packed, busy life, it's not easy. Laziness sets in. I compare it to physical exercise. If you want to be strong, you conquer laziness and do those push-ups. If you want to have friends, you make those calls. Like I said, simple.

People. People who need people. Are the neediest people in the world?

I veer to extremes in so many ways, but a major one is in my social needs. I can be completely alone for days at a time. If I'm focused on an art project, it's like I'm the only person on the planet. I consider myself to be an introvert. But put me around a table with my best friends, and I'm the loudest person in the room. Apart from my daughter and husband, only my besties can make my heart burst with joy.

I'm so grateful to have them for the laughs and love, but also because I know what it's like not to have any friends at all.

A Priest, a Crash Test Girl, and a Comedian Walk into a Bar . . .

When I travel to another city for PR or meetings, I get into being anonymous and alone, going out to eat and to movies. It's cheesy but true: I have a friend in myself.

But I also like to make friends with people for an hour or an evening, just to test my social skills. When I was in New York pitching this book to publishers, I went into my hotel bar by myself and struck up a conversation with a gay Catholic priest ten years my senior, and, on my other side, a twenty-five-year-old female improv comic from one of those *Saturday Night Live* feeder companies.

We just started talking, drinking, telling our stories, and bonding over our differences, and we became an unexpected trio of fast friends, staying at the bar for hours and laughing our asses off.

It was a bright, spontaneous human interaction that I look back on and smile. There are just so many people in the world, and it's as easy to talk to them as it is to avoid them. If I'm in the mood, I love nothing more than striking up a conversation, and making that freeing, anonymous, temporary connection.

HOW TO TALK
TO STRANGERS

- **Let it flow.** Allow the conversation to develop organically. If he or she gives you a polite smile, you might make a comment about the weather, the food, and then see where it goes from there.

- **Give to get.** Reveal a little about yourself, and they will respond in kind, and before you know it, you're fast friends.

- **Don't push.** Let people and conversation come to you, which they will if you smile. A smile is like turning on a light switch that signals "Friendly! Will exchange pleasant words!"

- **End it before it gets awkward.** Don't overstay. Classy move: Surreptitiously pay for their drink on the way out.

NO TIME FOR VAMPIRES

CONCLUSION

The best friends give you energy. They feed you, not feed off you. Their hearts and minds inspire you and make you feel like you can do more and be better. If you have anyone in your life now that drains the energy out of you with negativity, complaining, whining, backstabbing, get rid of them. Life is short.

STYLE

Whenever we planned an explosion on the show, we had a long discussion about which propellant to use: C-4, ANFO, detonating cord, propane, or gasoline for that Hollywood red, pillowy poof. They each had different functions and different visuals. After hundreds of explosions, we all knew which one was just right for the job.

However, we didn't limit ourselves to predictable choices. Once, we used nondairy creamer for an episode called "the Creamer Cannon." At the bomb range, we poured a pallet of nondairy creamer into a five-hundred-pound oil barrel, attached it to a pressurized tank to make it shoot into the air, and, simultaneously, ignited it to create the explosion. We usually used C-4 on the range, so the bomb tech (who, by the way, was just filling in for the day) wasn't too worried. He said, "Creamer? Don't worry about the blast shields. Just go stand over there. You'll be fine." We were about a hundred feet from the cannon and we thought we'd be okay. After all, it technically was really more of an ignition than an explosion. Semantics.

We triggered the pressurized vessel, sent the creamer into the air where it ignited into an unanticipated massive fireball. When the blast went off, I thought our bomb tech was going to pee his pants. The wind shifted, and it seemed like the fireball was a growling devil face coming right at us. I screamed, "Run!" Instinctually I grabbed Tory and used him as a human shield, while abandoning Grant completely. So much for altruism.

Coffee creamer rained down like sticky napalm all around the range, melting cameras, ladders, and our equipment. It smelled really good, though, like burned marshmallow. The sodium aluminosilicate (an ingredient to keep it from caking) in the creamer was what made it so flammable, much more than we, or any of the experts, predicted. It created the perfect thermic reaction for pyros like me, and we got it all with a high-speed camera.

Armageddon-size sugary fireball + comedic survival techniques + no injury = fun day.

Decisions, Decisions

Sometimes you're deciding which explosives to use. Other times, you're just picking out your OOTD. (I just learned what this acronym means. FYI, fellow analog oldies: It's "outfit of the day.") In both cases, your choice will have an effect. Even if you decided to purposefully reject everything that has to do with fashion and buy all your clothes in thrift stores, you are still making a style statement, such as, "Zero fucks given!" or "Thrift store chic." Even if you walked around completely nude, you would be telling people a lot about who you are based on what you are (not) wearing. We live in a society where a pink knit cap or a safety pin on your lapel are major statements. Your clothes and "look" are how you present yourself to the world. You supply the clues that will lead people to form a judgment about you, like it or not, believe it or not.

**HOW DO YOU
FIND YOUR "LOOK"?**

QUESTION

All Black, All the Time

MTV launched when I was in fifth grade, and I was obsessed. I'd leave it on all day long, and thought of pioneer VJs Martha Quinn and Nina Blackwood as my buddies. I remember dancing around the table to Michael Jackson and Madonna, and wearing lace gloves, which some might consider a fashion mistake, and braiding my hair to make it all kinky and crazy like the VJs. My somewhat misguided theory was hoping the VJs would help me define my own look and make a cool impression on other people.

My artsy punk aesthetic continued to evolve in high school and can be summed up as black-on-black, except for one crucial element: my hair.

Readers, I have a confession to make. I am not a natural redhead. My real color is honey blond, which is a nice shade, but it didn't play with my eighth-grade look of wild wannabe rock star clothes I sewed for myself. At fifteen, I went to the supermarket, bought a few packets of cherry Kool-Aid mix, and soaked my hair in it until it was bright red. This was way before the days of Manic Panic, and food dye was all I had to work with. I smelled like a bowl of cheap punch and would wake up with pink splotches on my pillow. When it was hot, pink sweat beads rolled down my neck, but it was worth it. As a redhead, like Rita Hayworth, one of my favorite screen goddesses, I couldn't be a shy mouse. I had to live up to my flamboyant hair, and the color change was what brought me out of my shell.

From shy to fly, thanks to red dye.

DRESS FOR THE PERSONALITY YOU WANT, NOT THE ONE YOU HAVE

I felt prettier, more confident, and unique. I was the only redhead (pink-head really) in our entire school, and no one else in my all-natural friend group would even think of changing their color. By making that one switch, I redefined myself—for myself and for others—and became, almost overnight, a rebel. The persona stuck, and I have never gone back. Over the years, I've tried variations, from strawberry blond to oxblood, once bright red with blond tips to look like a Te-quila Sunrise, blood red with black streaks, black hair with magenta streaks. I would say, "I'll try anything once," and still do.

Next came the tongue piercing, which I got at seventeen in the back of a hair salon, along with my then-and-now-BFF Brittany and a few guys in a metal band called Pink Toilet Paper. I don't remember who suggested doing it first, but we were all intrigued by the idea, so I walked in there and stuck out my tongue apprehensively. Like most teenagers, I wanted to stand out and be different, by doing exactly what my friends were doing. I wanted the adrenaline rush, the punk rock badass feeling, which I absolutely got. But by the time I got home, my tongue was pain-fully swollen. My mother said, "Why are you talking like that?" She was horrified to find out, and I tried to keep her from rushing me to the doctor. I kept it in for years, and even flashed some mouth metal in a *MythBusters* episode about electric fences. I touched my tongue bar on the fence to see what would happen. (The zap wasn't that bad.)

My style went with my rebel persona, and they reinforced each other. Throughout college, I curated my clothing daily. One day, I could have passed for a Metallica groupie and the next a rockabilly swing dancer. My favorite garment was a white leather trench coat. I paired that with red plastic pants, a concert jersey, jewelry purchased at the pet store for restraining large animals, and ornate black eye makeup. The pièce de résistance was a tattoo of a butterfly fairy on my lower hip that I got when I was twenty (now covered by orchids). I had the Happy Goth look *down*.

Since I was a broke student, I was a big fan of repurposing normal clothes into quirky masterpieces, making any old or boring thing look new and cool with a pair of scissors. I'd cut holes in the bottom of tube socks and wear them as sleeves (which were also useful in mopping up sweat at concerts, or as an oven mitt while baking). When I had the liquor ambassador job, I'd get tons of free promotion T-shirts. By cutting out the logo, I'd have strategically placed slits and hold the shirt together with a few safety pins. As a belt, I'd use a men's tie, and always wore a few dozen rubber bands on my wrists. Not only did I look punk rock, I was a walking multi-tool.

My only brief departure from my SF style was while traveling the world after college. The backpacker is a strange breed. They come in all types, but have a unifying aesthetic: Thai linen pants, sarong skirts, harem pants, Maori symbol necklaces, sandals, buns, tans, body hair, and matted not-necessarily-on-purpose dreadlocks. After a few months of traveling and trekking, I started falling in line. Backpacker brethren always recognize each other, maybe by our fashion code, or the smell. When we meet, we exchange our travel tips and short bios, and a round of one-upmanship of tales of travel trials, often featuring giardia of varying degrees.

In Jerusalem, I got my belly button pierced, which was a very small deal.

In Barcelona, I got my nipples pierced, which would have hurt more if not for the anesthetizing quality of local sangria. (FYI: Not recommended for anyone who intends to breastfeed. What a mess that was. It was like feeding my daughter with a sprinkler instead of a hose.)

After I returned to San Francisco with all my new hardware, it was a natural to return to my punk black-on-black, human multi-tool style, and I stuck with it for years to come. **Style does define you to others, and it helps you to figure out who you are, too.**

WHAT EVERY CRASH TEST GIRL NEEDS TO HAVE IN HER PURSE

I was at a wedding shower recently, and played along with the stupid games. One of them was for all the guests to empty their purses, and the person with the weirdest contents got a prize. All the guests upended their handbags. Out came lipsticks and electronics, glasses, wallets, hair accessories. And then there was me. I had a roll of duct tape, a test tube with a stopper, a multi-tool, a plastic replica of a grizzly claw (intended to use on the show about repairing a plane that had been mauled by a bear), oh, and a glass eye.

The look on their faces as they examined my stuff was priceless. One woman picked up the bear claw and said, "Who *are* you?"

Obviously, I won the lavender-scented lotion!

Not saying you should carry around glass eyes, bear claws, or test tubes, but there are a few things that experimental women should always keep in their bags, in case of a real, or a creative, emergency.

Army Surplus
gas mask bag

duct tape

multi-tool

pencil

pen

~~icrack~~
~~tech drug~~
phone

bandana

coin
purse

 gum

Band-Aids

 carabiner

 hand
sanitizer

 energy
bar

 tiny hot
sauce

~~safety~~
sunglasses

My style idiosyncrasies outed my inner rocker. When I dressed like a rebel, I felt like one. People believed me to be someone who would take risks and sure enough, I started taking them. Which came first, the adventurous spirit or the tongue piercing? My style and personality were inextricably linked. I felt good about myself when I dressed like a proto-punk. It was my look, and I was loyal to it.

Then, one day, I was riding in my car and stopped at a traffic light. I glanced at my reflection in the rearview mirror and realized with a shock that I looked like Avril Lavigne. I had socks on my arms, pigtails, six piercings. I started laughing out loud. *I'm thirty for fuck's sake.* I took off the socks and the piercings and shook out the pigtails all before the light turned green, and I never wore any of it again.

That stoplight was an indelible moment. I remember exactly how I felt, a bit embarrassed, a bit horrified, albeit introspectively amused, and deeply centered about making the change. By that point in my life, I didn't need any help defining myself. I knew who I was, and the message I wanted to project, and teenage pop star was not it. In high school, style had set me free. At thirty, it was time to retire my attention-seeking rebel wear and redefine my style to fit the road-tested woman I'd become, where it's about the woman, not the wardrobe.

CHOOSE THE RIGHT TOOL FOR THE JOB AND THE RIGHT OUTFIT FOR THE ROLE

EXPERIMENT

A wardrobe question unique to *MythBusters*: "What does one wear to blow up a building?"

Answer: Anything nonflammable and easy to run in.

SHOP GIRL

The word "wardrobe" sounds very professional, and something from another show. I think I speak for the whole crew when I say that the *MythBusters* style was, for the most part, utilitarian. Usually, I wore jeans and tees, but occasionally, I was supplied with experiment-specific gear, like a hard hat, safety goggles, leather welding aprons, odd one-off costumes like a silver boy-short bikini I wore while being painted head to toe in aluminum paint or special gloves for handling hypodermic needles in a condemned building. Though I did rock the hell out of some hazmat suits.

As for glamour, forget it. Professional hair and makeup people? Are you kidding? We scrubbed the workshop floors ourselves. I barely used any makeup on camera, because reality TV was actually reality, and who wears full matte makeup and lipstick to fit a pig stomach with plumbing parts? Rookie move. We showed up, did our thing while being filmed by a skeleton crew of our friends like making a home movie, and then we went home. Having a shiny nose or smeared mascara seemed trivial when scrubbing exploded poop off a wall or dodging a fireball. *MythBusters* was more lug wrench than lipstick.

In truth, I did try some makeup experiments of my own after seeing myself on TV. Lessons learned: Blue sparkle eyeliner looks stupid on camera, waterproof mascara fails when diving with sharks, and there is no makeup that can withstand a good on-screen vomit. Also, once the show went HD, concealer became your friend.

Wardrobe malfunctions were unique to the show: A visible bra strap in postproduction would trigger an automatic reshoot. (My rebellion was to go braless, boob-mando, but the joke was on me, because internet.) My bra underwire once stabbed me in the knocker when we naively became crash test dummies in an early episode about the airplane brace position. Also, while welding, I learned the hard way how many flammable synthetics are in women's stretchy jeans. And gun-friendly shirts are those with high enough necklines to cover your cleavage, thus preventing hot lead shells from landing between your boobs. Worst of all, shortly after having my baby, I peed my pants on camera during a timed test, but managed to completely hide it with a sweatshirt around my waist and finished the scene. The entirely child-free crew had not anticipated how much having a baby fucks up your bladder, and how challenging a long experiment with no breaks would be. (TMI? Welcome to being my new friend. Kinda how I roll.)

Fans noted on websites that my clothes did seem to be awfully tight. That wasn't a style choice per se or me trying to be sexy. If you were around machines with moving parts that seem to reach out and grab loose clothing, reel you in, and grind you into hamburger, you would wear tight T-shirts, too. Flouncy pants and peasant blouses could mean a gruesome injury. Early on, I wore scarves because the shop was freezing. Jamie would take one look at me and say, "Stop whatever you're doing and *take off that scarf.*" Old footage of me using the chop saw with a scarf loosely draped around my neck gives me the chills. You always had to be on your guard in the shop. As soon as you got comfortable—or warm—that's when accidents might happen.

After destroying most of my pants with sparks, grease, animal secretions, and the like, I started wearing enormous men's coveralls, because I couldn't find any in my size. I guess the majority of welders are not petite females—clearly an oversight since the best welders I've ever met were women (Scottie Chapman!). I had to go to specialty shops deep in the Mission and dig to the bottom of the pile to find something that actually fit. There aren't a lot of clothing lines for female construction workers, period, and designing a flattering silhouette was clearly not a priority. Comfort and functionality would have to do.

In the end, none of it mattered anyway. Over the course of shooting an episode, I'd be covered in grease and sweat, my hair would go limp, and my makeup would melt off. A small nose pimple becomes a live volcano on HDTV. We were exactly as gross and disgusting as we looked. The charm of *MythBusters* was the true grit of the cast and the show. To us, TV wasn't glitz and glamour. We cared more about how things were made, not how we looked. The aesthetics of the show fit me perfectly, or at least we grew into each other.

All that said, I like dresses! Practical comes first for me professionally, but I feel great about myself in classically girly clothes, too. Why can't a welder be a feminine woman? I enjoy blowing shit up *and* sparkly shoes. Not simultaneously. Usually.

In 2009, *MythBusters* was nominated for an Emmy for the first time. Instead of pulling an A-list move of asking a designer for a dress, I took the D-list approach and bought my own deep purple Gucci full-length gown. Unlike all the other women, I did *not* wear a push-up bra. That year, I'd just had a baby and wore a super-sexy nursing bra with three pads to sop up the leak from my sprinkler nipples.

MythBusters was a perennial nominee, and every year, I'd have my Cinderella moment at the awards show in a brand-new dress. I know it seems crazy to spend thousands of dollars on a dress that I would only wear once. Logically, designer couture is an impractical, ridiculous waste of money. I'm hoping that, someday, my daughter will ask for a prom dress and I can repurpose one of my vintage gowns for her. Or maybe I'll sell them to send her to college.

Regardless, walking the red carpet in full glam mode was an undeniable thrill that I hope to repeat one day. But what I enjoyed most about it was the inherent contradiction. In my pretty dress, standing tall in stilettos, people didn't recognize me. I'd watch them realize who I was, and their mouths and eyes would turn into big Os. I shocked reporters, photographers, and even my producers and friends by how well I cleaned up.

I can't say I was comfortable out there, neither at ease in the moment nor pain-free. I wore two pairs of Spanx and swallowed down a flock of nervous butterflies. It felt like my organs were shaking on the

This is what I looked like
the day before the Emmys.

You can take the girl out of the shop,
        but not the shop out of the girl.

CRASH TEST STYLE HACKS

- **Use toilet seat covers** to get rid of shine on your face for low-rent *papier poudré*.

- **Try desert dust** for a dry shampoo.

- **Use 600-grit sandpaper** as a nail file.

- **Rub the foam** that comes with dry-cleaning hangers to get deodorant stains off your clothes.

- **Shave your sweater** with a razor to get the pills off (like I did approximately five minutes before I did a segment on *Today* with Hoda and Kathie Lee).

- **Wear black polish** to hide grease under your nails and . . .

- **Wear black clothing** to hide dirt and grime.

- **Layer.** If one shirt gets covered in paint or grease, you can just peel it off.

- **Have many boot options.** I'm a boot collector and have different styles for all occasions: tall boots for snake country, waterproof boots for a muddy bomb range, booties for city walking. Because I wear tight clothes (see p. 106), and can't fit a lot in my pants pockets, I wear knee-high boots and slip my phone, cash, cards, and multi-tool into them for easy storage and access. It's perfect for shop work or an outdoor concert. Bonus: If your phone is in your boot, it won't fall out of your back pocket and into the port-o-potty.

- **Bandanas!** They work for sweat, pulling back hair, carrying large amounts of nuts and bolts, and as face masks for dust and the desert.

inside while I fought for "smooth TV star" on the outside. Still, though, I looked *good*, like I belonged. Only Tory and Grant knew just how nervous I was because I tend to giggle and smile and talk too much when terrified, which was perfect in that situation.

It was all a facade, and we were willing participants in it. *This is a fantasy, not anything close to reality*, I kept thinking, which was ironic because we were in the reality TV category. As a tongue-in-cheek moment to remind myself of the surreality of the Emmys, I always took a picture of myself on the toilet in my gown to send to Brittany every year. It was an inside joke between us, a social commentary on the absurdity and preciousness of the red carpet, how all these glamorous, impeccably groomed and styled stars were going to have to pee at some point over the long evening and night of festivities. **If you do have the occasion to glitz and glam it up, have fun with it, do NOT overspend on it, and remember that under all those sequins, you're still you.**

ANALYSIS

IDEALLY THE JOB YOU WANT IS A GOOD FIT, REGARDLESS OF THE STYLE REQUIREMENTS

When I was a liquor ambassador, I had to dress like a party girl. Not the job I wanted.

When I was a receptionist at the ad agency, I dressed in "respectable" slacks and sweaters, and hated my life.

When I hit the job jackpot and went to work at M5 Industries, and on TV as a *MythBusters* host, I wore what I would have put on anyway (minus the neckwear). For the most part, my "real" clothes were also my "work" clothes, so I guess I had always dressed for the job I wanted.

During the *MythBusters* years, I'd look at my filthy nails, finger Band-Aids and duct tape residue, ruined T-shirts and paint-splattered

I LIKE YOUR STYLE

I admire anyone who has great style, either fashionable or functional. In the fashionable category, I admire bravery and individuality and adore artistic expression in the stories of:

- **Iris Apfel.** My fave quote of hers: "Fashion you buy, style you possess."

- **Helena Bonham Carter.** Her unusual outfits and quirky sensibility give her the look of a real-life Tim Burton character. (They were a couple, FYI.)

- **Grace Jones.** No one embodied strength like her, or skewed gender aesthetics as fearlessly (with the exception of Bowie, maybe).

- **Deborah Harry.** She remains sexy at seventy-two because of her "don't give a fuck" attitude and self-confidence.

- **Courtney Love.** During the *Live Through This* era of grunge, she sold me forever on a silk slip and combat boots.

- **Basquiat.** His hand-painted clothing as art.

- **Kim Gordon.** My grunge nostalgia touchstone.

- **Lady Gaga.** For making fashion a performance.

- **Björk.** Because Björk!
- **Pharrell.** He can rock a park ranger hat, and always looks crisp and tailored within an inch of his life.
- **Prince.** The velvet boot-cut pants and frilly shirt king of all time.

In the functional category:

- **People who wear the same thing every day, like Steve Jobs and Mark Zuckerberg.** They don't want to waste time, energy, and attention on clothes, so they just wear one uniform every day.

- **Jamie Hyneman made his black beret and white shirt his signature look for years.** I asked him about it and he said, "Everyone remembers the beret guy."

- **Fran Lebowitz and Alfred Hitchcock.** Can't go wrong with the daily dark tailored suit; apparently it is a sign of creative genius.

- **Daisy Ridley.** As Rey in *Star Wars: The Force Awakens*, she wore her Jedi flats, which made sense since she was running around the woods fighting evil Kylo Ren. There shouldn't be heels in space.

- **French women.** From what I have seen, with wild flowing hair and a stylish romantic scarf, they age gracefully and just get sexier.

jeans, and think, *Dirty hands + dirty clothes = bliss.* I'd dreamed of a career where I'd be covered in paint and plaster, and when I got there, I was overjoyed.

I believe you do have to dress in appropriate attire for the job you're trying to get (if not the one you eventually want ten years from now, because who the hell knows what you'll be doing then?). Your first impression *should* impress, no matter what your profession is. When I have pitch meetings or auditions nowadays, I show up looking sharp and clean so they think, "She tried to look crisp, therefore, she cares and respects us." If I showed up looking like a slob (which wouldn't be "me" anyway), they'd think, "She can't be bothered, so why should we take her interest in us seriously?"

I recently had an important meeting at a major network, and you bet I agonized about what to wear. The show was about technology, and I knew the people in the meeting would be looking at me almost as a template. Could I pull off their vision of the tech-savvy role? I decided to keep it simple with nice black jeans, a clean gray shirt, a fitted jacket, and booties. The only real flair was earrings made from roller chain I stole from a robot.

How you dress reflects on you, but also shows your deference to the people you're dealing with. It's a nod, like saying, "I understand what you want from me." Once you get the job, all bets are off and you can wear whatever you want.

THERE'S A HAMMER IN EVERY TOOL

CONCLUSION

So how is style like deciding between gasoline, C-4, or nondairy creamer? You have to choose the right explosive for the given job to make the right impression (and know whether blast shields will be necessary). Every situation has a purpose, and the clothes are just tools to get the job done. The day before I walked the red carpet, I was running around the shop in a welding mask, and got that job done. On Emmy night, the job was to glitz it up for a surreal night.

Glamour moments when you obsess about every detail of your look are like desserts. They're great every once in a while, but too much of them will change you, not for the better. I like the balance of occasionally slipping into a fantasy world of the red carpet, but I live in my cozy home of comfort and functionality. I can be famous, sexy, and myself, in gowns or in coveralls. No matter what I'm wearing or how people perceive me, I'm still me, and there's more to me than I used to think.

In the shop, we can make anything work for us. Don't have the exact right tool? Use what you have however you can. It's the same with style. You don't need exactly the right piece to make or break an outfit. Make a great look with whatever you've got. It might be a challenge, but if it weren't, getting dressed would be boring. My *Myth-Busters* family is going to laugh when they see how I have appropriated our old shop saying for fashion:

CHAPTER SIX

ALCOHOL

t *MythBusters*, whenever we'd test myths about nau-
sea, we had to make the subject feel sick first before
we could experiment with possible cures. How to
make someone feel queasy in the shop? I invented a
vomit chair using a design I borrowed from NASA.
It was actually a dentist chair on a revolving platform. You sat in the
chair, and then tried to touch tennis balls attached to four points
around your head, while spinning. The dynamic movement is disori-
enting to the inner ear.

The first time we used the chair was to test seasickness cures.
Grant sat down and spun around until he felt on the verge of puking,
and then he'd eat some ginger. Or he'd sit, spin, nearly vomit, and take
Dramamine. Every day for a week, poor Grant had to get in the chair
and, more often than not, hurl his guts out. What a way to start your
workday. Isn't TV glamorous?

Not too long after, we did another test of hangover queasiness
cures. Every night for three nights, Tory and Grant drank themselves
crazy drunk. The first night was beer only. Then hard liquor only. Then
a mix of beer and liquor. They really put their bodies on the line, going
drink for drink until wasted. We calculated the amount it should take
by body weight, measured diligently and tested consistently. We built

a set so they could sleep in the shop, which was cold and mouse-infested. It wouldn't be a controlled experiment if they went home to their own beds, ate their own food, and puked in their own bathrooms. We monitored it all. Our reality TV was so much more real than most. People are always amazed how we actually did everything they saw on their screens.

The degree of pain in Tory's and Grant's hangovers was always worse if they'd had beer, alone or mixed with liquor. (I've found the same thing in my home testing, incidentally.) But since this was *Myth-Busters*, we needed measurable data for our conclusion, not just arbitrary anecdotal results. The test was about the intensity of the hangover, so the boys would have to push their queasiness quotient in . . . The Chair.

Grant was hurting bad when he staggered into the room, still drunk from the night before. He sat down, pale and sweaty.

I smiled at him, and said, "Just think. When you vomit, the test is over!"

And he immediately retched.

He said later that just looking at the chair, with all its bad memories from the seasickness cures test, was enough, and he couldn't hold back. Grant was so conditioned to associate the chair with puking, he hurled just from looking at it. He had an involuntary spontaneous reaction.

Sunday Funday

I have a framed poem by Charles Bukowski, author of *Barfly* and many other novels, hanging on the wall of my art studio called "The Laughing Heart." I wish I could reprint it for you, but I'd have to give the Bukowski estate my entire book advance. Here's one line: "You can't beat death but you can beat death in life, sometimes."

A famous overindulger, Bukowski might've been trying to drown death in life, most of the time. He was a literary hero of mine, and I admired his rebel spirit and respected his heavy drinking as his inspiration.

In my early twenties, before I started working on *MythBusters* and before I met Paul, I was reading alcoholic and junkie writers almost exclusively, and listening to addict grunge-era musicians. Oh, and drinking. Creating art seemed inextricably linked to drugs and booze, and, as an aspiring artist myself, I aspired to follow in my heroes' footsteps, living on the edge, and believing that great art would be the end result of copious consumption. I'd point to Nirvana and say, "Kurt Cobain is an addict, but listen to this song!" The people driving breakthroughs were using anything they could get their hands on.

I was really comfortable in dives all over San Francisco among the punk/grunge or rockabilly kids. There is something romantic about reading and writing from the end of a bar with my favorite scotch. The characters around me were great for both people-watching and inspiration. My roommates and I would roam from bar to bar in high-top sneakers, holey jeans, and flannel, play pool, shoot darts, and drink too much. As I've mentioned, I knew from my mom every crappy watering hole that offered free hors d'oeuvres (aka dinner), and would make the rounds to eat, drink, and create my persona as *that girl*, a wild child who lived each day to the fullest. We were just having fun, me and my equally artsy, equally grungy friends, crawling through the bar scene in our early twenties in a city known for its culture of freedom and expression.

One afternoon after I was at some dive with a bunch of friends, I came home to find my roommate Ariella angry with concern. "Kari. You have got to stop drinking. You are a mess," she said.

"But it's Sunday," I replied.

"Every day is Sunday for you!"

WHEN/WHERE/HOW MUCH SHOULD I DRINK?

Are You Drinking Too Much?

I don't know about you, but I have been wrestling with this question since I was a teenager, and continue to, even now that I'm a grown woman and responsible for another human life. I should not be having more than one glass of Chardonnay at lunch, but that one glass, at my local, artisanal pizzeria with the sun streaming in the window, really does feel like poetry. I have had all my most brilliant ideas after one glass of wine. I truly believe it is my perfect state of being.

My relationship with alcohol is long and complicated, because of my upbringing (more on that in a moment!), bouts with depression (more on that in chapter 8), social anxiety, and the fact that booze was, for a few years, my livelihood. No matter how much I justified my consumption, I kind of always knew it might be a bit much.

A friend of mine, a former smoker, got herself to quit by imagining every cigarette butt she'd ever flicked on the street or stubbed out in an ashtray appearing suddenly in a mountainous pile in the middle of her living room. Whenever she's tempted to smoke, she visualizes her own personal tobacco landfill. It turns her stomach and puts the kibosh on her craving.

This trick does not work for me. When I imagine every drink I've ever had poured into a swimming pool, I kind of want to jump right in.

The Fuzzy Navel

I come from a family of social drinkers and party lovers. Throughout my childhood and to this day, my parents host or go to parties five days a week. Actually, that might be a slow week for the Byrons. I

NOT MY FINEST HOUR

When drinking, I was full of reckless confidence and could lie through my teeth to my parents and take fantastic risks not to get caught coming home later, or going out when I wasn't supposed to. Besides the basics of always having gum in my backpack, I would formulate answers to their questions—"Where have you been?," "What have you been doing?," and "Who were you with?"—before I walked through the door. I recently told my mom that I used to sneak out of the house through a window, or tiptoe through the front door, and either take the car out in the middle of the night, even before I had my driver's license, or jump onto my bicycle.

She had no idea. "You *did*? That's terrible!"

It really was, because I rode my bike home loaded. I still shudder and thank my lucky stars that I didn't get in any accidents.

grew up eating hors d'oeuvres for dinner, as in, "Really, Mom? Bean dip, *again*?" Just kidding. I love bean dip for dinner!

All of our family traditions had deep roots in alcohol. On Christmas, we had Irish coffee in the morning and Bloody Marys in red plastic cups at the afternoon parade. On Thanksgiving, we drank wine all day. On Tuesdays, my dad and grandfather always had Manhattans. They gave me the cherries. A coming-of-age rite of passage in the Byron household: When someone in the family turned twenty-one,

ZOMBIE APOCALYPSE
CRASH TEST KIT

1. Remove foil and cage.
2. Hold bottle by punt.
3. Run machete along seam with smooth continuous glide right through the lip with the force of slamming a book.
4. Drink champagne for courage.
5. Now you have a machete for decapitating zombies and a sharp bottle to penetrate brains through the eye socket.

↑
weapon

zombie ⟶

she or he got what we called a "Mae West." We had special hourglass-shaped glasses that emulated the curvy female body of the famous foulmouthed actress (love her). Grampa poured 7UP into the bottom chamber, and "smooth as silk" Kessler Whiskey, aka rocket fuel, into the top. Downing it in one gulp was like a shot and a chaser in one glass. "Puts hair on your chest," the men would say. (Not much of an enticement for me.) I associated celebrations with getting tipsy, and we celebrated *a lot*.

In the 1990s, my parents ran a champagne cellar every weekend, hosting tastings and giving tours of the cellars. Basically, they worked for wine. Did you know the sound of clinking glasses is called "tintinnabulation," meant to sound like the ringing of church bells and to ward away evil spirits? I do, because I have heard my dad share this little fact almost a million and three times. He and Mom had a deep appreciation for the beauty of wine—the color, the aroma, the ritual of the pour. Could they be any more Northern Californian? **In our family, it wasn't only that a drink could make you the life of the party. There was no party without it.**

My drinking life apart from family parties began in earnest in high school and it served a very specific purpose: Drink to open up. From what I'd observed at my parents' parties and from my own experiences, people got loose after a drink or two. Drinking was the perfect remedy for all my high school social problems.

Before a kegger in someone's backyard or basement, I'd pre-game with my friends who were a bit older and knew the special liquor store just outside of town limits that didn't check ID. Our inaugural teenage drink of choice was a Fuzzy Navel, the perfect cocktail for a teenager's taste buds, consisting of orange juice and peach schnapps. After three of them, I felt like the most charming person in the room. While my friends used alcohol to lighten up, I used it to transform. I knew, from the bottom of my heart (and glass), that if I hadn't had a pitcher's worth of Fuzzy Navels, I wouldn't have been talking, laughing, smiling, making eye contact with cute boys, or playing spin the bottle.

I would have been cringing in the corner. **There's a reason it's called liquid courage, and drinking Absolut(ely) worked for me.**

**DRINKING HELPS YOU
BECOME A BETTER
VERSION OF YOURSELF**

The Crutch

At college and later while traveling the world, I grabbed a drink whenever, wherever I felt insecure or anxious. I'd drop into unfamiliar situations and environments every several days while traveling, and all that newness triggered my anxiety. Fortunately, my tried-and-true remedy was always available at a moment's notice.

In those years of meeting new people constantly, alcohol helped me break the ice and strike up conversations. A drink melted away my fear—and theirs—and I could talk to anyone, about anything, and be charming AF (I thought) while doing it. I used to say, **keep your lovers close but your bartender closer.**

While I was traveling, every night was a party. After the second glass, when most people around me were closing out and asking for waters, I kept going. Most mornings I woke up with a hangover. Many nights, I didn't sleep at all. I've had chronic insomnia since I was a kid. I used alcohol to shut down my racing mind.

In Egypt, I tried drinking heavily to make myself pass out, and, instead, stayed up all night painting a cityscape on the wall of my hotel room in a crazy fit of insomnia. Hotel management wasn't as pleased with the mural as I was. I had to run out of there, lying down in the back of a local taxi chased by cops with machine guns. Of all the countries I visited, my most insane experience was in Australia, a country and people known for hard drinking. I got so wasted there, I put on a party dress and knee socks and drunk sleepwalked through a flood. And the next day? I was right back at the bar, ordering a hair of the dingo, mate.

TO LIVE LIFE TO THE FULLEST, HEAD OVER TO THE BAR

EXPERIMENT

From Australia then back to San Francisco, I'd gotten on a track of trying to live each day to the fullest—as Bukowski would say, "beating death in life"—and that meant drinking. A lot.

He's far from the only writer whose work is alcohol-soaked. Famous wit Dorothy Parker was an alcoholic, as were Hunter S. Thompson, Truman Capote, Stephen King, F. Scott Fitzgerald, Ernest Hemingway, Carson McCullers, Jack Kerouac, and the list goes on endlessly. Drinking seemed like a requirement for literary achievement, and added to the romance of a writer's life, something about pouring martinis and one's soul onto the page (speaking of martinis, Ian Fleming, James Bond's creator, shook, not stirred, massive amounts of them).

Oscar Wilde, a heavy champagne drinker, once wrote, "Work is the curse of the drinking class," but it might've been even more apt to say, "Drink is the curse of the writing class." They just go together, like peanut butter and jelly, peas and carrots, rum and Coke.

I wasn't driving creativity with drinking anymore. I was just drinking. I wasn't creating great art, just wasting time. I'd wake up and think, *What should I do today? Guess I'll head over to the bar . . .* It was a habit. What was that Dean Martin quote? "I feel sorry for people who don't drink. When they wake up in the morning, that's as good as they're going to feel all day."

I'd love to tell you a rock-bottom story, if I could remember it. I'm sure it happened, but I was blacked out at the time. I did get kicked out of a bar once because I bit a biker dude. He was getting inappropriate, so I chomped on his arm. The last thing I remember was the taste of leather.

125

Lil Q *About Alcohol, De-Mythstified*

We did *so many* shows about alcohol that attempted to answer:

Are beer goggles a thing? For this test, I rated photos of thirty random men on a scale of one to ten while sober, tipsy, and drunk. The idea being, when drunk, I'd give the men higher ranks than I would have when straight. What I realized was that, when sober, confident-looking guys were attractive to me, but when drunk, they seemed cocky which was a turnoff, but the vulnerable, sad guys looked appealing. My score was kind of strange. I was harshest about the photos when tipsy, liked some guys when straight, and others when drunk. I'd say that, for me, the beer goggles test was inconclusive, but plausible.

Is drunk driving more dangerous than drowsy driving? To test this, Tory and I drove on a closed course when sober, after downing a few shots, and after staying up for thirty hours straight. We were both better drivers well rested and straight, no surprise there. The drunk test showed marked impairment behind the wheel, but being exhausted was far worse.

In my case, I made three times the mistakes compared to drunk driving. Tory was ten times as bad. So the answer is: Don't drive when drunk or drowsy!

What works to sober you up? For this test, the guys did five shots, and then tried a handful of sober-up cures—black coffee, a slap in the face, ice water dunk, vigorous exercise—after which they tried hand-eye coordination tests used by astronauts of drawing along a zigzagging line for accuracy and speed. According to the data, coffee and an ice water dunk? Did nothing. But five minutes of exercise or receiving a slap did show some improvement. So next time you need to sober up fast, have someone slap you really hard, and then run in circles until you fall down. Just a suggestion.

One off-camera alcohol myth: My producer and I were being eaten alive by sand fleas while we were working on shark myths at a lab in Bimini. I asked Jamie how he got away unscathed. "Gin, inside and out." Not the answer I expected. He rubbed it on his legs and drank some. I tried it and I swear they started to leave me alone! Not really science but an anecdotal sample of two. Too bad *MythBusters* isn't making any new episodes. That would be a fun one to test.

Ariella was one of my three roommates at the time, in a small but lively apartment. We all lived like this, and I was taken aback that she'd singled me out and told me flat out to stop. In my mind, I was only keeping up with *her*.

I said, "Thanks for your concern, but I'm fine."

Her words stuck with me though, and, as angry as I was, I did take heed. I was so mad because, well, she was right.

I slowed it down and stopped going to the bar for a minute. Shortly after that, I met Paul, and he became my focus. I had this wonderful boyfriend, so why did I need to go to the bar every day and night? Paul and I would party together, too, but nothing like I'd been doing on my own or with my friends. It was more fun to be at home with him, drawing, anyway. Maybe I grew out of it or maybe I realized I really never needed it.

Truth is, I *was* a bit of a disaster. I thought I was so funny and charming but in reality, I was a bit embarrassing. (Thank my lucky stars there weren't phone cameras back then!) Luckily everyone around me at the time was a disaster, too. **Drinking to inspire great art might work for some, but for me, it was counterproductive.**

ANALYSIS

INSTEAD OF HELPING, ALCOHOL WAS COUNTERPRODUCTIVE TO ART, LOVE, AND FRIENDSHIP

If my saturated period resulted in museum-quality sculpture and publish-worthy poetry, I might still be drinking. But all those days and nights of partying undermined my creativity. *Not* partying was when I made good stuff, like my travel journals or the drawings and art Paul and I made for and with each other.

LETTERS TO A YOUNG POET

I slowed down, thanks to my friendship with Lisa, and a parting gift she gave me when we went our separate ways after Egypt. She opened a little silk bag she'd sewn and took out a book, Rainer Maria Rilke's *Letters to a Young Poet*. This strange little book of advice spoke to me philosophically about my need for others to bring me happiness, finding poetry, and going into yourself for answers. Before too long, without making a conscious decision, I started drinking a lot less. Lisa is the one who encouraged me to start seriously journaling, to pour my anxiety and depression into my art, and mitigate my insecurity and sadness that way. Emotions, all of them, even the dark ones, were material I could mine for my poetry and drawings. Instead of dulling feelings with booze, I could experience them, and turn them into art.

When I came back to San Francisco, I wasn't sober by any means, but I had an alternative remedy at my disposal. I could make myself feel stronger by reaching for a pencil instead of a glass.

WORKING WITH
DRUNK PEOPLE

When I started working as a liquor brand ambassador in my midtwenties, I'd spend every night in bars handing out free drink samples. My theory was that it'd be a fun job because everyone I met would be buzzed and feeling fine.

I got a totally new perspective on alcohol by not drinking. At the events, I'd witness normal humans devolve into disheveled animals who slarved over each other after a few free drinks. When you're sober, you see the ugly truth, that hammered people are kind of horrible and gross. I got the offer for the unpaid M5 Industries internship at the same time I could have advanced in a liquor-related career. One of the reasons it was so easy to say "no" to the corporate booze job was the idea that I'd have to hang out with drunk people all the time. Not for me. **Drinking at work might've been a good idea at this job, because it wouldn't have revealed the ugly truth to me.**

If I had an addiction to alcohol, there was no way I could have stayed sober at all those events. Considering how turned off I was watching drunk people, I realized I was the opposite of an addict. I had a man I loved, and a new job I was excited about where drinking was absolutely not allowed. We were operating heavy machinery and one drink would have put any number of people at risk. It was even dangerous to show up hungover due to lingering impairment. Nope, drinking was not part of my future career (unless I got drunk for science), so I just slowed it all the way down. It wasn't hard at all. Don't think I could have gotten through the grosser experiments with a hangover.

I did come out of that time with something: embarrassing home movies of parties that made me look nothing like the cute, suave bon vivant I thought I was. I thought I was the coolest, smoothest person there, but the evidence says otherwise. I will always be grateful that I exhausted my experimental years before social media. For me, I can use my experiences as personal lessons. They only exist in the mythology of my life, not on the internet, where they'd remain *forever*.

Waking up with a savage hangover many mornings for, oh, about ten years, wasn't good for my health, either. I was young and thought I was bulletproof, but I must have known that drinking had consequences. The sleeplessness. Drinking dinners. In those days, I weighed around 105 pounds. It would take hours for me to get over the morning sick stomach, pounding head, and the shakes before I felt close to normal. The icebreaker, the little bit of help, had turned into a crutch. I handicapped myself to the point that I couldn't function without a drink in my hand. **Using booze as an icebreaker broke me.**

Samplering

It might be a bit optimistic to say that after I met Paul, I just naturally grew out of drinking. It remained a constant battle in my head. Was drinking a celebration or a crutch? Was a glass of wine a simple pleasure or a source of guilt and shame? Alcohol had been a part of the fabric of my life since childhood. I knew too much of it wasn't good for me, but I liked it.

Much like Grant puking at the first sight of the chair, I drank at the first sight of discomfort. While drinking wasn't "involuntary" for me by any means, it was certainly knee-jerk. I'd conditioned myself to drink whenever and to whatever quantity necessary to satisfy my current state: anxious, sad, happy, insecure, hungover. It was a habit that at times felt like something I did without conscious thought.

For example, I would reach for a glass under the following conditions:

- With family.

- With friends.

- At any party or celebration.

- When meeting new people.

- If I felt sad.

- When I couldn't sleep.

- To inspire creativity.

It goes without saying that if you imbibe every day in such amounts that you wind up biting a biker on a Sunday afternoon, you're drinking too much. I had to sit myself down at several points in my life and say, "This might be a problem." I also noticed that, if I couldn't see it myself, there were friends who sent up a warning signal every time. Lisa encouraged me to draw instead of drink to chase away the blues. Ariella challenged my drinking rituals during my romanticizing period. Chasing art and dreams with Paul was more fulfilling than glass-filling.

And then, there's Stella. During my pregnancy, I was dry for months on end, which I hadn't been since the age of fifteen. I realized I liked the control of sobriety. After she was born, the idea of being out of control of myself in front of her, or even away from her, turned my stomach faster than Grant in the vomit chair. No way was I going to be incapacitated if she might need me.

I still have wine, I love it. I am a Northern California girl, and wine is in the blood. I can never turn down a good scotch. But I've found that when I drink too much, it can trigger depression, feel terrible the day after, and make it harder to be a good mom.

I guess alcohol is like a frenemy. **She can be so much fun and I adore her, but damn, if she can't make your life harder when she feels like being a bitch.**

CONCLUSION

TO PARAPHRASE
WILLY WONKA: A LITTLE
NONSENSE NOW AND
THEN IS RELISHED BY
THE WISEST WOMEN

CHAPTER SEVEN

SEXUALITY

n one of our last seasons, we took on the myth called "Is Bigger Better?" about whether large boobs led to bigger tips. Try going into that one with an open mind. To test it, I wore DD size breast prosthetics while working in a café. No problem about making coffee. I'd been a barista in a former life for years. However, my experience with wearing giant silicone cutlets in my bra was nonexistent.

The most fun part of the episode was shopping with Tory at the Piedmont Boutique in the Haight—where sequined bikinis, wild costume dresses, feather boas, sassy wigs, and prosthetic boobs aren't consigned to any particular gender or size. Otherwise, it was an adventure in awkward. As soon as the word "boob" came up in an ideas meeting, everyone in the room immediately turned in my direction. Of course, I'd be the one who had to do the testing.

Funny, I can't remember our doing a myth about "Is Bigger Better?" regarding penis size . . . *how* did we miss that one? I guess we're lucky the show got canceled. This particular myth came up toward the end of the series, and my first thought was, *Uh oh, we're scraping the bottom of the barrel now.* We always said we'd never run out of myths, but were we really doing this one? Seriously?

The critical thinker in me asked, *How can we prove this? How can the data collection be consistent? How will I wear giant hunks of silicone all day and not get a backache?* During any of the rare uncomfortable tests, I changed the language of my thoughts from "I have to go to work today" to "I get to go to work today." From "Come *on*" to "Get into it." I tricked myself into being positive. If I'd fixated on the leering nature of this episode, I wouldn't have been able to do it, but I figured out how to make it interesting for me. I'm usually very naive about the male gaze. According to my friends, I never know when a guy is checking me out. This gave me an opportunity to look for it.

Shocker: Dudes tipped 30 percent more when I wore XL hooters than when I wrapped myself up for the "small" size, or didn't do anything for the "control" size. What you don't see in the results on the show is that my natural boobs were tipped the least of all three options! My own body was the least profitable. Should I be insulted? LOL. (Can't believe I just wrote "LOL." I spend way too much time on the internet; I have lost all language to convey sarcasm or irony without an emoji or hashtag. #wtf.)

That day, we proved the obvious about making tips, but I also think I proved that by taking ownership of your role in something, you can do things on your own terms in an empowered way. We all had to do some crazy experiments. But I had to represent all of womanhood every time we did a myth about gender. I felt it was important for me to put in my two cents and make sure the myths, questions, and discussion didn't slant too male. For example, we discussed doing a test on the topic of "throwing like a girl." I hate all phrases that include "like a girl" ("drive like a girl," "run like a girl," etc.), because they are always derogatory. They imply weakness and incompetence. The guys in that room didn't mean to be negative, and the point of taking on that myth was to discredit the idea that women were bad at throwing. Didn't matter to me. To me, the whole premise would only reinforce the derogatory prejudice. The entire conversation was like chewing on tinfoil. I said, "As the female representative for the world, I have to object to the way we're presenting this myth." I had to speak up, for my sake, as the one who'd have to do the throwing or driving or running. But also for the sake of the show, so it didn't skew so male that it alienated our female audience.

I worked hard to be part of the crew and not "the girl on the show." But since I was, there was no escaping it. Nearly every day on *MythBusters*, I was reminded of the line about Ginger Rogers, "She did everything Fred Astaire did, but backwards and in high heels."

QUESTION

CAN YOU BE A SEXUAL PERSON, A STRONG INDEPENDENT WOMAN, AND A FEMINIST AT THE SAME TIME?

Busted

A page-one Google image search picture of me is a photoshop of my head on a bodybuilder's body in a white tank top that says BUSTED. (Double entendre, anyone?) It's been on the internet for so long, people think it's real, even though it's obviously not me. Fake news? How about fake nudes.

Speaking of photoshop, I've also seen some weird, fake porn of myself. You can watch me stretch in slow motion thanks to someone out there uploading a series of screenshots of me wearing a gray T-shirt (a T-shirt!) from one episode. In fact, if you were so inclined, you could see stills and videos of me in any compromising position from the show: fixing the microphone in my tank top, running, or bending over to any degree for any reason. I'm sorry, did I say "compromising"? I meant, just doing my job. When I'm on-screen, I basically can't move my hand up and down in any linear manner, or a hand-job-y GIF is sure to follow. I learned the hard way not to suck the end of a party balloon. That one was too easy for a cyber troll fan to resist.

That said, I'm a grown woman with a sexual appetite and a wealth of knowledge and experience. I like attention. I love my body and the way it makes me feel. Like most grown human beings, I've come a long way in my sexuality. There was a time when I was a confused virgin no one wanted to sit near, let alone have sex with. It's a bit of a catch-22 to have gone through the trouble of developing and learning the power of my sexuality, only to be objectified for it later on.

LOSE IT

CRASH TEST

I lost my virginity at eighteen, methodically. I had a plan: A good friend's older brother had a reputation for being promiscuous, and he was a drummer so I knew he was an easy mark. He was in college, and I was a senior in high school. The entire encounter was blessedly brief. It was completely devoid of anything that you could describe as special, passionate, or loving. We had spent the evening at the Sick and Twisted animation festival downing Budweisers from a backpack. (That's right. *Beavis and Butt-Head* as foreplay.) We were both drunk, and all I remember about that night was a lot of awkwardness. He had no idea it was my first time. He thought it was a casual hookup. Poor guy.

He asked me years later, "Why did you let me do that?"

I told him, "I kind of used you."

I wanted to get that task out of the way before college: 1) Fill out college applications, 2) Get a summer job, 3) De-virginate. I certainly could have waited for the fairy tale ideal of having sex for the first time with your one true love, but I was deeply suspicious of fairy tales. (Why would princesses always have to have some sort of injury or near-death experience to get the prince? Makes no sense.) I went for it because true love wasn't happening for me, and once I was rid of the burden of inexperience, I could relax and find love without the stigma of virginity.

Problem was, nothing really changed. I was still the same girl. The constant lesson of adolescence, one you need to learn again and again, is that you aren't defined by others. **ASSERTION put me in the driver's seat of my sexuality, but it didn't take me anywhere I really wanted to go.**

Honestly, I had no idea about the power of sex. The nonevent of losing my virginity did inspire me to learn more, and, for many years, I was a very active student.

THE ONLY WAY TO LEARN ABOUT SEXUALITY IS TO SAMPLE, AND SAMPLE WIDELY

HYPOTHESIS

Try on Many Skins

My college years and early twenties were a wild time of random hook-ups and walks of shame, not an uncommon story. I was attracted to dangerous-seeming men, the outside-the-box kind of guys with tattoos and punk haircuts who were into nefarious things, like pool hustling and maybe pot dealing. I've ridden on the back of many motorcycles to and from skeevy dive bars and made out with "not boyfriend material" bikers. Musicians, artists, actors, and poets were my favorites. If a dude came off like a broke, brooding outsider? He got my attention. If he seemed responsible with excellent career prospects, he'd get nowhere with me. It was my rebellion from my vanilla suburban small town.

Yuck. So cliché!

I had an all-too-common pathology: Good girl likes bad boys. What made them so irresistible? I convinced myself that outsider slackers were "honest" and "brave." Bad boys radiate confidence, and I sought what I didn't have in them.

Musicians were the best. I had one of the sexiest times of my life when I hung around with my two best guy friends from that period of time. Both had that heroin chic, starving artist, just-got-out-of-a-recording-session look. Somehow, their consummate passion for music translated into an overt sexuality. It radiated from them like a shock wave. The three of us would lie around in a dark room, candles flickering and Jane's Addiction on the CD player. I was in my Courtney Love phase in silk black slips with smudged black eyeliner and combat boots. I lounged between them—we were always fully dressed—and they each ran the tips of their fingers up and down my arms and legs and over my body for what seemed like hours. That was it. It never went farther. It was the big tease, the thrill of the chase. We knew the excitement was in the potential, the kinetic energy. If anything actually happened, it would have ruined the friendship. Instead, it was like a powder keg near a fire.

You can't judge a lover by his cover. A guy who dresses like a criminal might be a feminist intellectual. I found that the toughest looking were the most tender. From these two undercover good boys in bad boys' clothing, I learned that the potentiality of sex could be hotter than the actuality. Not even kissing was the hottest sex I ever had.

I read feminist and erotic authors like Simone de Beauvoir and Anaïs Nin, and they stoked me to experiment further. I made out with a girl. Not so shocking now (thanks, Katy Perry), but it raised some eyebrows in the '90s. It seemed rebellious, and therefore, I liked it. In hindsight, my desire for this girl wasn't sexual. I wanted to *be* her, not make out with her. She was a talented artist who looked like Debbie Harry. We were drinking friends and, one night, we were hanging out with one of those sexy musician boys and it's possible I kissed her so he would be impressed. But then he left, and I thought, *Let's try that again, just to see if I liked it.* Experimentation requires sufficient data. **By trying on many skins and attempting to define myself as a sexual person, I learned what I was—and wasn't—into.**

It's my opinion that sexuality is on a spectrum and we're all sliding around on it. We waste a lot of precious time worrying about where we fall. Instead, we should all just be who we want to be. I kissed a lot of different people to learn about myself and what I liked. With a large sample, you get better results. I was very careful to be respectful and let them all know they weren't part of a long-term study.

One guy seemed like my perfect match. He checked every box for my ideal boyfriend: good looking, a writer, a good talker, a musician. He drove a rumbling muscle car. On paper, he looked like my ace. He adored me, but something was slightly off. He didn't have the right scent. It wasn't offensive, but I just didn't care for it. It was so confusing. I wasn't physically attracted to him, but I tried real hard to be, and kept dating him anyway.

When I met Paul, I realized boxes don't mean shit. Attraction had been a mystery for so long, and then it was solved with our first kiss. The smell of him, his look, his voice, his eyes, I felt pure animal attraction. Sex with him was powerful *and* easy. Even more powerful, Paul and I were in sync out of bed as well. During my promiscuous years, being smart and funny took a back seat to seduction. But with Paul, I realized the sexual potency of making someone want to talk to me, not just jump my bones. **My wild days were fun but not fulfilling.**

Sexuality without feeling was falsely empowering. Being a rebel in this department didn't add up to much. I fell in love with Paul, and finally understood what great sex was, because we wanted so much more from each other (and it didn't hurt that he is totally sexy).

IF YOU'VE GOT IT, FLAUNT IT

EXPERIMENT

Sexual Symbolism

Almost immediately after the first episode of *MythBusters* (the one featuring my butt) aired, I became an unwitting, unlikely sex symbol. My behind went viral, and my sexuality was no longer in my hands.

What to expect if you become a lust object of the nerdiverse?

- Fan sites devoted to your T-shirts.
- Fan pages of screenshots of your feet.
- GIFs of you greasing a shaft.
- A porno based on your show, starring someone who is supposed to look like you.
- Facebook pages for your butt.

I wasn't *trying* to be a sexy TV host. I was just being myself. If anything, I was going for asexual in my overalls and ponytails. But to our audience, I was a cute, upbeat, accessible girl on their screens, having fun doing stuff they cared about, and a meme was born.

The internet is populated by cretins with too much time on their hands and just enough technical know-how to try to humiliate any woman that triggers their rabid misogyny. My attitude is, I didn't make the GIFs or porn. I'm not responsible for them. They exist, but I have nothing to do with it, and, therefore, I can't let it get to me. You can't fight "rule 34." (Not a nerd? Rule 34: "If it exists, there is porn of it.")

I don't take the GIFs seriously, or chafe against the objectification. Other people get incensed on my behalf, however. I heard Jamie's wife was annoyed with how I was portrayed. (Of course, Jamie is married to a rad feminist. Love her.) Maybe my skin thickened as the years passed, but I'd just shrug and say, "Whatcha gonna do?"

Don't get me wrong: **Sexual objectification is inherently sexist.** But it's not like I could stop the internet. Getting upset didn't serve me.

This kind of thing happens to any female public figure (and many nonpublic figures). When the first female Dr. Who in the show's fifty-plus-year history was announced, I set my stopwatch to see how long it would take for nude photos of her to appear online. (One day. What took so long?)

YOU ARE BOTH FULLY IN CONTROL OF YOUR SEXUALITY, AND POWERLESS TO ITS INTERPRETATION FROM OTHERS

The Fantasy

About ten years ago, *FHM*, a men's magazine, asked to photograph me doing a science experiment in my underwear and high heels. I would have preferred *Vogue* or *Glamour*, but they didn't call. I was in a rebellious period, looking for a kind of performance art piece on sexuality and realness, and so after some consideration, I wanted to say "yes." But a woman in power at the Discovery Channel heard about it, and called me and asked me not to do it.

"It's a huge mistake," she said. "No one will ever take you seriously again if you do this."

I was shaken by her plea and almost reconsidered. But as "rebel" was my default mode, I suddenly wanted to do it even more. I saw this photo shoot as fighting against the fact that a woman couldn't be a tomboy welder, *and* smart, *and* pretty. Saying "I'm serious" and "I'm sexy" should not be mutually exclusive. When the Discovery exec told me that posing in my underwear would diminish me, it felt like another weight the world put on me to hold me down. I know the executive was trying to help me, but she actually convinced me to do it. I defined myself, no one else. That's what feminism means to me.

I went to New York and worked with a team of professionals, the photographer and hair and makeup people, in a clean, classy studio. The sophisticated environment of the shoot did not reflect the magazine itself, which could only be described as "super cheeze."

It wasn't a *total* exploitation. I wore a lab coat over my bra and showed only a little cleavage. I wear less at the beach. Discovery sent along Amy, aka "the Keeper of the Button." Her sole job at the shoot was to make sure that my lab coat stayed closed. The photographer would ask me to undo the one button, and Amy would jump in and say, "I don't *think* so."

The fantasy they were going for was something like, "The hair comes down, the glasses come off, and lookie here at the sexy scientist with the Bunsen burner!" I spent most of the shoot trying not to laugh. I enjoyed the day immensely. It was hilarious to pretend to do science in the ridiculous outfit and everyone there was in on the joke.

I collected a new experience. It's what I do.

When I saw the raw pictures, I liked how I looked and felt excellent about their realness. Then, the photographer said they were going to photoshop the image to make me even *better*. I watched with slack-jawed shock as they ballooned my boobs and thinned my thighs and arms. The proportions suddenly looked absurd. I get why they blew up my chest, but did they have to shrink my guns? It's sexy to look too weak to pick up a fork? Really?

At the time, I thought the photoshopping was just a demonstration—one tech artist showing off for another—but they ran the altered images anyway. When the magazine came out, I laughed at how absurd it was. When I showed it to everyone at M5, they reacted the same way. Tory said, "Who is this Ginger Lab Barbie they're calling Kari Byron? And where did those boobs come from?"

Fun fact: The day before I left for this shoot I tried one of those "don't try this at home" magic tricks where you make a fireball with the gas from a lighter and I accidentally burned off all my eyelashes. So the lashes in the photo were fake, too.

In the end, I learned that objectification is about me, but it's not me. The sex symbol and the actual woman are not one and the same. How could I feel personally objectified by an image that erased my realness? It might as well have been Jessica Rabbit. Talk about dehumanizing. The magazine was trying to sell an unrealistic "ideal" to readers who might then expect real women to look like that, which is a shame. The raw photos were so much sexier. I was proud of how I looked in them. I would have preferred they appeared in print and represented a real woman with a real body. **To feel sexy in your realness, unbutton your figurative lab coat and show yourself to the world.**

Sexuality is a line to walk. You want to feel the power of it. You want to be wanted, but not to be wanted only for that. *FHM* magazine might have understood its readers, but they didn't get the fans of *MythBusters* at all. The vast majority of them liked me for my personality and accessibility, not because I was a fake-boobed bimbo. The audience appreciated me as a full package, a real live woman who talks, chews gum, and fires a rocket at the same time.

I *am* the geek girl next door.

I'm as real as it gets, big arms and all.

CREATE HUMAN LIFE

CRASH TEST

Interestingly, my next big photo shoot was for *Pregnancy* magazine, and featured five photos of me holding a giant pickle, rocking in a chair, wearing a full-coverage trench coat, looking extremely real— and enormously pregnant. No one at *MythBusters* told me that I looked hot in those.

You don't often see pregnant women on television. Usually, when an actress is pregnant, she stands behind waist-high counters or carries a gigantic tote bag to hide the bump. Only on reality TV will you find women letting their Buddha bellies enter the screen before they do. Throughout my on-air pregnancy, women wrote to me, thanking me for being a visible working mother-to-be. I was grateful for the letters, but the truth is, I didn't have a choice. We were in the middle of the shooting season when I popped. I guess the timing wasn't ideal. Paul

and I were actively trying, but we didn't expect it to happen the first month.

My first thought upon finding out I was knocked up: *Yeah, baby!* My second thought: *Well, that's the end of my career.* How the hell is pregnancy going to work on *MythBusters*? I was super worried that the cast, execs, and crew would see me as a liability. Other women could wait until the last possible minute to share the news with their bosses. I couldn't hold off on telling mine because my job would have to change. I couldn't lift heavy objects, eat weird things, be painted with aluminum, or be around noxious gasses. I couldn't participate in certain dangerous situations—speed driving or hang gliding, for example—that I'd ordinarily jump at.

It actually turned out that my having to ask for help was a blessing for everyone. The producers saw how quickly we could get things done if Tory, Grant, and I weren't solely responsible for cleaning up the whole shop after a shoot. Even after years on a hit show, we were still doing the grunt work.

I used to be one of the guys, but now there was no denying I was a woman, with egg-making ovaries and an occupied uterus. Every day at work during those nine months, I crash tested pregnancy on Discovery. I was the first woman in the history of science entertainment TV to be visibly gestating on air. (FYI: Nine months is a myth. It's really ten months or forty weeks.)

If a man is expecting a baby, he is not defined by it. People don't assume his schedule, focus, or commitment to the job will change. But when a woman is expecting, she has to contend with a new label: MOM. No one asks male colleagues, "Are you going to quit after the baby is born?" But a lot of people on the show and in my social circles asked me, and were surprised when I said, "Why the hell would I do that?" Pregnant women don't actually transform into wood nymphs in flowing linen dresses and flower crowns who hum lullabies at their engorged bellies. I was pregnant, but I was still me.

BUMPS, BRUISES,
BITES, AND BURNS

I sustained countless injuries on *MythBusters* and other shows, all on behalf of science, and I've got the scars to prove it.

- **Stomach burns.** For an episode of *Head Rush*, against my better judgment, we were boiling water in a covered Erlenmeyer flask. Bending down to look, I said to the producer, "With nowhere for the steam to vent, aren't we creating a pressure vessel, a bomb?" Just as I stood upright, the beaker exploded. Boiling water and glass blew into my stomach. I lifted my shirt up to find blistering second-degree burns on my skin. Two seconds earlier and it would have hit me in the face. I went to the hospital and got treated, but to this day, I have a scar that looks like a wolf when I get tan.

- **Metal flakes in the eye.** Once, while using a grinding tool on metal, a flake of it got through the side of my safety goggles and embedded in my eye. I had to have it removed by a doctor, because metal splinters rust in the eyeball and have to be drilled out.

- **Backfire bruises.** Every time I fired a rifle, I'd get huge bruises on my shoulders.

- **Lost toenails.** I once dropped a chunk of steel on my boot, which bruised my pinky toe so badly, my nail turned black and fell off.

- **Bite marks.** We did a myth about whether a guard dog could bite an arm off. I wore a protective suit, but when the dog bit me on the arm, I got a canine-mouth-shaped black-and-blue mark that didn't fade for weeks.

- **Internal rash.** I didn't know I was allergic to polyurethane until I inhaled fumes and found I was affected systemically. A rash would pop up randomly on the surface of my skin as red, tiny bumps itchier than poison oak.

- **Hair removal via head mold.** We made a mold of my head, and the release agent didn't take, so when Tory pulled the casting off of my face, most of my eyelashes and eyebrows came with it. That hurt. The hairs are still embedded in it. I had to be very creative with makeup and false lashes for a while there.

I hadn't changed, but the people around me sure did.

There was even a weird backlash from male fans who didn't like to see their accessible nerd girl next door turn into an expectant mother. Their perception about me changed, and my hotness cooled. The irony killed me: My sexiness quotient dropped when the result of my active sex life became apparent. Another irony: Pregnancy sex was *the best*. I've been chasing that dragon ever since. It's an evil joke that you have the best orgasms in your life when you feel the least sexy.

The fan site threads about my pregnancy were not kind. But when I came back after my maternity leave, they lit up again about how quickly I "got my body back." I hadn't realized it'd been gone. Of course, no man deals with this scrutiny. The sexism around my pregnancy bothered me far more than the fetish sites about my feet. Having Stella was the greatest thing I'd ever done, but my post-pregnancy weight loss, according to some fans, was a more worthy accomplishment. You had a baby? Who cares. Looking sexy again was all that mattered to them.

The only way to normalize pregnancy is for more women on TV to get out from behind counters and tote bags.

They have to put their bellies front and center, and show that pregnancy is just a normal state of body, a thing that occasionally happens. I love the character of Laurie on HBO's *Silicon Valley*, who barely even acknowledges that she's pregnant while starting her own company. One of my heroes is stand-up comedian and writer Ali Wong. Her show *Baby Cobra* (view on Netflix) was filmed when she was seven months pregnant. She was up there, all over the stage, on all fours, cracking jokes about feminism, marriage, motherhood, killing it—while gorgeous. It's like mothers are supposed to be sexless June Cleavers, or hot-and-horny MILFs who seduce the pizza delivery guy, but neither stereotype generally exists.

Is Bigger Better?

Sexuality is in the eye of the beholder, and that eye is in your own head. In the *MythBusters* boob experiment, people gave bigger tips to a curvy barista, and we learned that some aspects of sexuality are predictable. But when you go a fraction of an inch below the surface, sexuality is unpredictable. You can't control what makes you appear sexy to someone, beyond obvious cues, which, let's face it, won't sustain a relationship for long. If women hope to progress at all, we have to be comfortable striking our own balance, even if you don't get bigger tips or a men's magazine changes you to make you "sexier" than you are naturally.

On the other hand, burying your sexuality to be taken seriously is no solution. It's limiting and gives a society that fears and marginalizes women's sexuality permission to define you.

IF YOU CAN'T GET OUT OF IT, GET INTO IT

CONCLUSION

I see sexuality as another tool in my belt, the one I use to make the world bend to my will. It's okay to use your sexuality intentionally, on your terms. I know for a fact that if I smile and play to a man's ego, he'll give me a chance. You wouldn't be serving yourself if you didn't use whatever you had to get your foot in the door. For me, it comes down to being myself, and being in control of how I present myself.

I wouldn't have been on *MythBusters* if not for my feminine charms. I'm aware of that, and just as aware that my shapely behind would only get me so far. You have to stay the long hours, do the hard work. I was a vegetarian, but I scraped chicken guts off the wall and ate bugs. I had to prove that I deserved to be there, *despite* having a cute butt. My creative contributions had to be excellent without fail.

But, if I didn't want to, say, move something too heavy, I could get the job done with a machine or use a pipe for leverage. In some cases, a smile was my leverage.

My women's studies professor in college used to say that strippers could be feminists who used their sexuality to serve their own purpose. I wrestled with the concept, and learned to apply it to my own life. How to contain the beast of male desire and use it for your own benefit? There are risks. You might lead the wrong man a little too far. If you get a job using too many wiles, your bosses and colleagues might resent you, and you'll have your work cut out for you to earn their respect.

Let's face it: Being a man is still a huge advantage. Women are underrepresented in science TV, government, science, tech, engineering, math, robotics. Even the "woke" corporations in tech are only 11 to 14 percent female. If it comes down to a man vs. a woman, the man will get the job, and he'll offered more money to do it, too.

A long time ago, I realized that the only way I was going to change the world—at least my small world of science entertainment TV—was from the inside out. The only way I could be "one of the only women" or "one of the first women" was to get in the door in the first place and dig my claws into the role. My female butt broke through that firewall. Then, once I was there, I had to prove myself. I did it

153

again, and again, every single week. By the time *MythBusters* ended, I wasn't the "first" and "only" female host in my field. There were a handful of us, and I was proud to have had something to do with the presence of women in science TV. I hope in some small way, being the token girl made a difference for others. In the final analysis, that's a net gain.

You might not be able to get out of an inherently sexist culture at work, which is awful and uncomfortable, but you can use critical thinking to turn a bad situation into a crash test. What can you get out of it? What can you learn from it? How can you use it to your advantage? Ride those waves, no matter how over your head they seem.

CHAPTER
EIGHT

DEPRESSION

n *White Rabbit Project*, we did an episode about mind control. We were trying to replicate the super-power using technology to control another human's brain. The first step was to test on a cockroach, a really huge one that I had to touch with my hands. After anesthetizing it in ice, I affixed a tiny circuit board to its back, and put electrodes in each of its antennae. When he woke up, he'd been transformed into a cyborg. I could control my roboroach's movement by sending signals from my iPhone to the electrodes in his antennae, and direct him around the table where I wanted him to go.

The next step was to use some of the same technology on a human subject. I invited Tory to dinner at an Italian restaurant and had my new neuroscientist friend hook up both of us with electrode pads on our arms and jaws, and, through a small computer, link them to each other. When I signaled my brain to move my arm, the signal went through the electrodes and into Tory's arms, forcing him to make the same movements I did. I had control and could contract his muscles, and he couldn't stop it from happening either.

So, when I asked Tory to pour the wine, I jerked my arm, making his spasm, and the wine got all over his clean white shirt. When the spaghetti and meatballs came out, I made Tory fling them all over the table, along with his fork. He couldn't take a bite if I flexed my jaw, and the sauce dribbled down his chin. Why do I love torturing him so much? I was snort laughing at how much fun it was to exert mind control on my human puppet. And the more wine I drank, the more sadistic this experiment became. It was the most fun I had on that show.

Tory? Not so into it.

We called this "mind control" but it was technically just me in control of his reflexes. Real-life mind control is a harder experiment. Nothing has taken my brain hostage like the reality of depression. That is the real mind control.

Depressed

During the *MythBusters* years, my role was to be the happy girl, the one who could be relied on to laugh and squeal at gross things. I was doing plenty of hard-core work, but always with a smile and a giggle. No one on the show or in my wider social circles knew a secret I'd kept for some thirty years: I contend with severe bouts of depression. They started when I was a kid, and still affect me now to a lesser degree. The most demanding aspect of being on camera for me is smiling through depression. It's been *way* tougher than forced vomiting or sawing through bone.

For most of my life, I never told anyone about my problem for a bunch of reasons. I blamed myself for feeling bad, and didn't want to burden anyone else, which would have only made it worse. I was afraid that people would stigmatize me as weak or defective. My mood wasn't anyone else's business. It was my private hell, and I made a conscious choice to wrestle with it on my own. It was my crazy to control.

The late, great Carrie Fisher, one of my all-time heroes, once said, "Take your broken heart and turn it into art." We all know how she struggled with mental health issues and substance abuse right up until her tragic heart attack, but she talked about it, she wrote about it, and she resisted it. She was the leader of the resistance in so many ways! At the Women's March the day after Trump's inauguration day, I cheered

for all the signs with Princess Leia's face that read, JOIN THE RESIS-
TANCE. If only she got to live long enough to see that. I wish I believed
in heaven so that Carrie Fisher could see herself become a symbol for
strong women who will not be silenced or shamed for anything their
bodies or minds might do.

In the spirit of Carrie, I want to share the story of my experiences
with the illness, which, I promise, will not be too depressing to read.

HOW DO YOU DEAL WITH DEPRESSION?

White Knuckles

At twelve, I met depression for the first time.

It was in the '90s when people weren't *depressed* per se. They were
just *sad*. You "got the blues," or "were on the rag." Everyone knew that
girls lost their minds at that time of the month. In hindsight, I think
my depression *was* triggered by sudden adolescent hormonal surges,
and has been a part of my life ever since.

I knew what I was experiencing was beyond run-of-the-mill teen
angst. It felt like falling into a deep maw of self-doubt. A crippling
heaviness pressed me into my bed all day, and my mind went feral
with dark thoughts: *You're worthless. You have no friends. You're no
value to anyone.* I visualized the depression as a dark purple stone inside my heart that was like a lead weight. I prayed (in my atheist way)
that my big red heart would, eventually, crush the stone heart, and I'd
go back to normal.

In the meantime, my strategy was to wait it out. I distracted myself with escapes that didn't require me to get out of bed. After school,
I would go to my room, lock the door, and crawl into bed like a
wounded animal to read a book or watch TV. *Three's Company* reruns
or *21 Jump Street* provided some relief.

157

If my parents wondered why I spent so much time in my room, they didn't ask. I had no idea what or why the sudden depression was happening, and the mystery probably exacerbated the condition. I might've Googled my symptoms, diagnosed myself, and read about what to do, but the internet didn't exist back then. I was in the dark about the darkness. I did have an inkling about what was going on, having seen glimpses of it before when my grandmother and my father would disappear into black moods. My grandmother lived to be eighty-seven, so the "heaviness" wasn't fatal. Dad always bounced back, too, and his example strengthened my resolve to white-knuckle through.

I could have talked to my mom, a woman who seemed to know everything medical, but I was too ashamed. In my family, we swallowed negative emotions and presented a happy front. I was afraid that if I told my parents the truth, they'd be destroyed. With them working so hard to take care of my sister and me, they had enough to worry about. So I kept my mouth shut and got really good at hiding my symptoms.

Sometimes, the self-loathing would grip me for only a few days. Sometimes, it would last for weeks with wavering intensity. When it was *really* awful, I would set my alarm to wake up at 5:00 a.m. so that I could watch reruns of *Little House on the Prairie*, and lose myself in the simple pioneer life for a couple hours. I wanted to be Laura Ingalls, that spunky little optimist, so badly. Channeling her powered me out of bed and got me through school for the day. By the time I returned home, I was exhausted by hours of pretending to be normal.

The same patterns continued through high school and college, when I lived in dorms or in apartments. When I crawled into my cave, my roommates probably just thought I was "in a mood" that lasted for weeks. They were focused on their own lives, and gave me space. Eventually, I'd snap out of it and be the perky party girl again.

While traveling, I could be in the most beautiful place in the world, and holed up in some grubby hostel by myself, wallowing. If I

158

clued in to the fact that I was missing out on the wonders around me, I'd force myself outside and start walking. Alleviating depression was like that Santa song, put one foot in front of the other. I would just start walking, and miles would pass. The scene would slide over me like fabric in the wind, and fly right over my head. I'd be miles away before I realized how far I'd gone. My knees would be achy and I could barely breathe, but I felt better. I walked through it.

I'd talk, too. Walk and talk. I would go up to strangers and just start chatting with them, pretend like I wasn't depressed, and they never guessed that I was. By making myself look someone in the eye and listen to his words and process what he was saying, I was able to connect back into the real world. Without TV, I turned people into my new escape. I remember once meeting a couple of artists from Paris in Prague who were posting LOST KITTEN signs. They would take flyers from other countries and make copies and post them far from home. Somewhat mysteriously, it was a statement about isolationism (I love the French; life is art even when there is no gallery or audience besides the artist). Their project became my own. I jumped into it and started to help. Making a statement on isolationism while traveling alone and battling depression in a far-off land was beautifully symbolic. Putting up flyers got me out of my head, and that was a huge relief. Eventually, the heaviness did fade.

WHITE-KNUCKLING WORKS, BUT AS A STRATEGY, IT IS PAINFUL AND PASSIVE

HYPOTHESIS

I reveled in the recovery, though. Once the darkness lifted, I'd feel euphoric, mitigated by the knowledge that this, too, would pass. My entire adolescence felt like walking a tightrope. I was unsteady socially, physically, emotionally, and mentally. My inner world was constantly changing, and I had to learn to adapt or die.

I wish I'd tried to get help, if not from my parents, then from a counselor at school or a doctor. I think of that twelve-year-old girl riding out the storm alone in her bed with the covers up, and my heart just breaks. I didn't know any better at thirteen, sixteen, or twenty. I endured it for years with nothing but grit and patience—and by being a very good actor.

JUST TAKE A PILL! YOU'LL BE FINE

EXPERIMENT

Girl meets Pill: I finally talked to a doctor when I was in my late twenties, and got a diagnosis. Previously, I had always blamed my depressions on outside sources—being broke or in a fight with someone—because I couldn't figure out why I was crushingly sad. Well, for the last episode, I had nothing to point the finger at. I was in love, had a great apartment, a dream job, and I was in perfect health. I had to face reality: The problem was in me. The doctor prescribed Wellbutrin and Celexa. I thought I must have it bad if I had to take pills morning *and* night.

Girl takes Pill: Like a good patient, I took the pills exactly as prescribed. The effects didn't kick in for two weeks, and when they did, I hated them. They robbed me of the good stuff, like my artistic impulse and hunger. Plus, the pills caused disturbing heart palpitations. I clenched my jaw until my dentist gave me a mouth guard. I kept with the regimen because I wasn't crying. I wasn't feeling *anything*.

END MY CHEMICAL ROMANCE

Girl loses Pill: I tried and tried. My chemical romance lasted on and off for longer than it should have. I had a great life and a dream job and I didn't want to screw that up with my episodes of sadness, so I put in my best effort to fix it with a magic pill. I experimented with different ones. But they all made me feel like I wasn't myself. **Pills did not make me happy, or solve my problems. They just created new ones.** In the end, I stopped taking them. I didn't want to be sad but I also didn't want to lose the realness of life.

The End.

The Horror, the Horror

The first years on *MythBusters* were both thrilling (I loved the work and, for the most part, the people) and disheartening. One of the bosses was a sadistic producer who got off on humiliating and tormenting the staff with pranks and practical jokes that, believe me, were *so* not funny.

Once, I was at the shop, sitting on a wooden table, working on a project, and he poured a powerful superglue and then an accelerator on my butt where I was sitting. It created a fast thermic chemical reaction that was so hot, I screamed. He thought it would be so funny to glue me to the table, but it was reckless and burned me. My underwear and jeans were glued to my burned skin. In the bathroom, I had to tear the cloth and a layer of skin to get my jeans off. I walked around for the rest of the day with my pants held together with duct tape.

Hilarious, right?

I didn't complain, though. At the time, my attitude was "Screw you, asshole, you don't get to see how you affect me." I dried my pain and humiliation tears, and kept a straight face around him. The way I retained my dignity, I thought, was to never let him know he got to me.

Secretly, he was known as Colonel Kurtz after Marlon Brando's insane despotic character in *Apocalypse Now*. As a woman, I was a particular target for his "jokes," but he humiliated everyone. He just kept upping the level of his sadism, moving the line of what was considered acceptable behavior, and we got used to it. *Awful* became the acceptable norm. We chalked up his bullying antics—like pantsing a new hire on his first day in front of the entire team, underwear and all—as just another of his plague of "jokes." The others at work sympathized with the victims, and would say, "I'm sorry he did that to you," but we were all too intimidated to complain.

The problem was, he was both in power and incredibly intelligent. He always made you believe that you were the weak one or overreacting if you got upset. According to him, it was just good ol' workplace hazing. He was regarded as a creative genius, a huge asset, and he was far more important than the little people, even the ones who appeared on camera. He was in charge of us, and his bosses weren't accessible to us. They were just too far up the totem pole. It always bothered me that they knew and would "talk" to him about it but never really do anything or create any consequences.

In hindsight, I wish I'd fought back. I was young and vulnerable, and he was the boss. Maybe if I'd done more, or rallied everyone to unite against him, we could have crash tested a way to fight the power. I was just too fearful I'd lose my job and I didn't have the experience or confidence to realize that this wasn't how the workplace was supposed to be. It was my first TV job, and, for all I knew, it was typical for the boss to give his employees shit, torment them, and cause physical harm. I wasn't the only one who let him get away with murder. Whenever Tory, Grant, and I talk about Colonel Kurtz now, we shake our heads in disbelief that we put up with him. We think we all had Stockholm syndrome.

The stress, fear, and powerlessness triggered a deep depression. Since the source wasn't going away, I envisioned misery without end.

MARRIED WITH DEPRESSION

The worst thing to say to any depressed person: "Cheer up! It's not so bad!"

Depression isn't a choice. Why would I choose to be miserable on a beautiful day, with everything going for me, in love with an amazing man?

Paul's challenge was to understand that depression happened to me, and to him, but it was neither my, nor his, "fault" and there wasn't a damn thing he could do about it. Of course, he was frustrated, and often blindsided. Depression isn't predictable. It can strike when you least expect it, and it's never convenient.

His learning curve was steep. He's an arty weirdo, too, with his share of bad history, but he'd never been *this close* to depression until me. I wear my heart on my sleeve with him, so he'd see me at my worst, no holding back or hiding. He was surprised and a little scared when I would go on and on about what a hot mess I was. After a lifetime of not sharing with my parents or anyone, I was too honest with him. We had some fights and struggled to accept each other, good, bad, sad, and ugly. It took years, but we got there.

How to Deal with Depressive People:

- **Make this your mantra: "It's not about me."** Do not take a depression personally. It's most likely not about something you said or did. It's not going to stop because of something you do or say. Just be patient and loving.

- **Take care of yourself.** You have to deal with depression, but it doesn't have to drag you all the way down, too. Do what makes you happy so that you'll have some good vibes on reserve.

When I got home from work, I was a basket case. Paul would drive me to a field of flowers where I would cry it out for hours. It was hard on me, but equally hard on Paul. He was just as powerless to help me, which was stressful and frightening for him.

Paul encouraged me to try pills again, this time, for stress relief. He saw me in pain and reached for the tools we had at hand. If I couldn't figure out how to calm down about the boss, I might lose my chance on the show, a consequence I'd fight tooth and nail to avoid. So I agreed to go down the psychopharmacological road again. Even though I was freaked out by the long list of the pill's side effects—too vivid nightmares, suicidal thoughts, weight loss, weight gain—it seemed worth a shot. It felt like I had to take the pill to keep my job. If that meant being an employed zombie, so be it.

For millions of people, pills are excellent tools to combat depression, stress, and anxiety. If drugs work for you, great. For me, they were deadening, again. I felt like the light went out. My jaw muscles were so tight, they ached. Electrical currents seemed to rocket through my body and shock me over and over. I hated the second round of pills even more than the first, and stopped taking them after only a few weeks. It was a long hike out of the Valley of the Dolls. Coming off them, even after a short stint, was a nightmare of cold sweats, anxiety, and insomnia. After a rocky couple of months to get back to baseline, I realized that our "Pills: The Sequel" crash test was flawed at inception. Work stress caused the depression. Instead of treating the depression, I should have focused on eliminating the stress.

In that work environment, it wasn't so easy. I did whatever I could to avoid the bad boss, and used other de-stressors at my disposal, like exercise, healthy eating, and spending time with Paul and our friends to remind me what I had to be happy about. I was able to get my anxiety and stress under control, and felt like I'd taken my power back by proactively addressing my reaction to the problem, if not eradicating the problem itself.

PROACTIVITY, NOT PILLS

ANALYSIS

The epilogue on Colonel Kurtz: Karma (not Martin Sheen) came to get him in the end. He burned too many bridges and was eventually asked to leave—and then claimed to be the victim of circumstance, not the bully he actually was, proving the theory that all tormentors are just insecure people who lash out at others to make themselves feel better about themselves. Ironically, his actions hurt him more than any of us, and now he's basically blacklisted from working in TV. It took much longer than I expected, but he burned his reputation from one network to the next. Now, he's mostly an online troll. Can't wait for him to pan this book. It will only expose him if he does, though. To paraphrase Carly Simon, "You're so vain, you probably think this chapter is about you."

GIVE BIRTH

CRASH TEST

Bear with me. This crash test has a backstory.

I'd been at *MythBusters* for several years. Kurtz was long gone. Things were great at work, and we were busting bigger and more complicated myths every week. We decided to re-create a luxury car commercial. One car raced across a desert. Another car was held aloft by a helicopter, and then dropped. The idea was that the car on the ground would hit the mark at the same time as the dropped car to prove how fast it was. I used physics to figure out the terminal velocities for dropping a car from the sky on a horizontal or vertical configuration (in relation to the desert floor).

On test day, I was in the Big Dawg helicopter, a Sikorsky Sky-crane, with the drop car on a quick release hook, hovering over an X on the Mojave Desert floor. Grant and Tory were in another helicopter, remote-controlling the land car. I was looking at the landscape, in awe of it, and with the fact that this was my *job*. I had one of those wild moments when it all hit me, the craziness of what I did for a living, how much I loved my coconspirators, and how cool it all was. I had to take a breath to get my head back in the game.

Our timing was precise, and we executed. At the right fraction of a second, I released the drop. The car fell. The other car came roaring across the desert, and they hit the mark at (close to) the same time. It was a huge success. I started screaming in my chopper, so overjoyed. When we landed, I didn't even wait for the cameras. I jumped out of the helicopter and ran over to the boys. They ran over to me. We started jumping up and down, screaming, hugging in a circle, celebrating like little kids.

The essence of our job was to play and have fun and do these crazy things. We had only one shot this time (we wrecked our one test car), and we pulled it off. We were ecstatic. Later on, we went back to the hotel with the crew and ordered dinner and drinks. I remember sitting in the hot tub, drinking a beer, eating pizza, thinking, *I can't imagine life getting any better.* It was beyond anything I'd imagined and I wanted it to stay exactly like this.

The next day, I had an eight-hour drive home in a truck full of equipment with my cameraman Scott. I was watching the *Sex and the City* movie on my laptop (don't judge) to make the time pass. In a scene, Big gives Carrie a pair of shoes, and for some reason, that little love gesture made me start bawling uncontrollably, the kind of sobbing when you snot yourself.

Scott just sat there and listened to me sobbing and sniveling over Carrie Bradshaw's shoes. He didn't say anything and must have thought I'd come unhinged. I didn't understand what triggered the emotion. It wasn't depression, I knew that. So what was it? A thought popped into my head: *What if I'm pregnant?*

The night before, in the hot tub, I'd turned to Tory and offhand-edly asked, "Wouldn't it be messed up right now if I were pregnant?" My unconscious is a lot smarter than I am.

I got home and bought a pregnancy test, which confirmed my suspicions. I was completely knocked up. No wonder emotions were so high doing that experiment. I'd been running through the high heat of the desert like a kid, toasting to the happiness of my life right now, and that long day culminated with the news that life was going to change *dramatically.*

I told the cast and crew while we were sitting around the office. People were telling real off-color jokes, as usual (TV productions are pretty lowbrow). Someone said, "Good thing there are no kids in the room."

I said, "There might be one."

The heads swiveled in my direction, everyone saying, "*Whaaat?*"

I remember nodding and kind of shrugging. A mother on *Myth-Busters*? How was this supposed to work? I guess I have a long history of making huge life decisions quickly and just going for it, and then problem-solving the adventure as I move along.

I chose my doctor specifically because she was clinical and not nur-turey. I wanted straight answers, not sympathetic looks. My friend Lisa had gone to her and asked a million questions about food, and autism. The doctor looked at her and said, "You're not that special. The odds are, you're going to have a healthy baby." I thought, *Yes, I want that lady.*

I had some questions for her myself, like, "So, when do I have to stop shooting guns? Is it when they get, like, little ears?"

She said, "I don't know. I'll look into it."

I ended up searching for policewomen with kids, and interviewed them to get my answer: At around five months, you had to stop shooting guns due to the noise and you had to be very conscious of the potential lead exposure.

Another question for the doctor: "If I've got twenty-five pounds of C-4, what distance do I have to be so that the shock wave won't affect the baby?"

She said, "What did you say you do for a living again?"

I basically had to figure all this stuff out myself (what else is new?). At around three months, I started to feel baby kicks during filming coinciding with my adrenaline rushes. At five months, my bulletproof vest got real tight. During one show, as I zipped up my flak jacket over my huge belly, the voiceover said, "It's a little late for protection now, Kari!" Emmy-worthy writing, right there. It didn't make the final edit of the show. It should have!

Throughout pregnancy, I didn't have a single depressive episode. It was happiness on tap. Everyone on the show was thrilled for me, too. For many of them, my pregnancy was the first one they'd been exposed to. It was a fascinating novelty to all of us, and we made our own fun with it.

Around five or six months, I smelled beef stew at a restaurant and had to have it. I went from being a lifetime vegetarian, to eating beef stew for a week straight. Maybe I had an iron deficiency? I went into work and told everyone about it, and suddenly they all broke out in groaning or cheers. That's when I learned they had an office pool about when I'd break down and eat meat. I was also their test case about how fast a belly grows and whether a mother's stamina and personality were affected by pregnancy. Baby-making is creating and building at its most elemental. They all wanted to share in the joyous experience.

I worked until the week of my due date, forty weeks pregnant on a bomb range, getting chased by an out-of-control robot. The robot was pretty slow, but I was really big, so it was a fair match.

I took four months off *MythBusters* after Stella Ruby was born. My first day back, I shot a fifty-caliber gun off of a cliff into a runaway car and hit the engine block dead-on, stopping the car.

It was a fantastic "I'm back, bitches!" moment, but it wore off by the time I got to the car to drive home. I could barely get excited about it. I had severe postpartum depression—headaches, extreme fatigue, crying jags. It started soon after giving birth, and was later compounded by my stressful back-to-work schedule.

Along with filming all week for *MythBusters*, I had a spin-off on the Science Channel called *Head Rush* on Saturdays. I'd get up at 5:00 a.m. to drive to my parents' house and drop off my daughter, and then race to be back for an 8:00 a.m. shoot. I was also breastfeeding and pumping throughout. I'd put on makeup to cover my dark circles, stuff all the stress, sadness, and anxiety into a compartment in my mind, smile, and dive into filming. I was a zombie.

I remember talking to a producer and opening up to her about what was going on. "What can I do? I can't stop crying. How do I smile when I want to crawl into a ball?" I asked.

She'd worked with news anchors who'd had to keep it together while covering wars and atrocities, on top of whatever they had going on in their personal lives. She looked me in the eye and said, "You just do it. You dry your tears, clean yourself up, and keep telling yourself that you're not you on camera."

And I did it. I just fucking did it. But by the end of the workday, I was a wreck. I'd worked through migraines, flus, and broken knees, but this was worse. On TV, you don't have a choice. It wasn't just *my* paycheck at stake. The twenty people behind the scenes working around me needed me to "show up." I remember swallowing my hurt and forcing myself to get perky. It was like pretending to be happy in high school, except with cameras and a thousand times the pressure. Occasionally, I locked myself in a bathroom or outhouse to cry for a minute. Then I dusted myself off, and faked it.

After five months of it, I was at my wit's end. Postpartum depression proved to me to be the grand mommy of them all, and she was a huge bitch.

I was holding on by a thread but this job was bigger than just me now. The guys talked me down from the ledge a few times. I stuck it out, reminding myself that it would pass, but it was sure taking its motherfucking time. In hindsight, I think I knew that if I didn't go to work, engage in challenges, be active, and learn new things, my depression would have been even worse.

169

SCIENCE THE PROBLEM
TO SOLVE IT

Wisdom and maturity have allowed me to be better equipped to deal with my depression. I learned the hard way what works for me by experience and experimentation. I can't just throw off the weight, but I can ride it out and push it away faster if I follow my four-step protocol that minimizes my symptoms.

Step One: Healthify. Drinking copious amounts of alcohol and eating junk food doesn't help, duh. I eat a healthy whole foods diet and avoid wine or Ambien, anything that's a downer.

Step Two: Exercise. Science tells me that endorphins released from sweating help. I do yoga, Pilates, or go on a bike ride every day. Sometimes the real hurdle is to just get out the door when all I want is to sleep and watch TV. Going to a class makes me socially accountable. It's bad form to leave in the middle of one, even if you're crying. I'm forced to stay for the hour and work my body and give my mind a break. I have ridden my bike until the tears dry in salty streaks across my face. Sometimes I will go down a hill that scares me, so I have to get out of my head and concentrate on not crashing. My Juliana mountain bike and a yoga mat have saved me countless times.

Step Three: Socialize. Surround yourself with people and talk to them. To be social, you have to get dressed and sit upright around a table and tune into the conversation. It gets you out of your head. At work, I have to put on makeup and smile in front of a camera or crowd. Sooner or later, my brain catches up to my smile. I also rely on my dog for interaction. I get companionship, love, and exercise in one furry, cuddly package.

Step Four: Create. Give those clammy hands something to do. I've sculpted through sadness (and happiness) and made some heartfelt works. You can see the darkness in some of my pieces from certain time periods and, interestingly, those are the ones people always ask me about and are drawn to. They have become reminders for me of rechanneling emotion into art. I had these feelings, they were turned into this piece, and now it's on the shelf.

My mood disorder has been a hindrance, no question. I won't sugarcoat that. But, like all my faults, I had to learn not to blame myself or hate myself for having it. There is a side of my life that's unpredictable, frightening, and dangerous. But so is life, so in a way, I'm innately prepared. It made a big difference when I stopped seeing my issues as burdens and recast them as problems in need of solutions—and I've gotten really good at finding inventive solutions. I mean, it's what I do.

I psyched myself up by thinking about Stella. I told myself, "I just had a baby girl, and I've got to do this for her. She needs to see me as a brave, hardworking woman who doesn't give up!" I learned in due time that my daughter didn't care that I was working so hard for her. She just loved me—and I was head over heels in love with her. My particular type of postpartum depression didn't include ambivalence or hostility toward my baby. I adored her, and that love was the light at the end of the long, dark tunnel. I never had a doubt that I wanted to be a working mother. I just didn't realize it would be this hard.

After six months, my put-on-a-happy-face became genuine, and the depression started to lift.

It was gradual, but even small improvement was huge. What a relief it was when my biggest working mom problems were logistical only, like finding a convenient spot to pump on a bomb range or in the desert, and where to store the bottles. I put them in the cooler where the crew kept their drinks. One guy (talking to you, Matt) accidentally spilled breast milk on his hand, and he ran out screaming like he'd been doused with toxic waste. When I heard myself belly laughing at the sight, I realized I was genuinely happy and the dark period was really over.

I fell back into the camaraderie at work. I brought in a batch of homemade cookies and the crew inhaled them as usual. I said, "Isn't it so cool that I baked all those cookies using my *breast milk*?!" and then I watched them all dry heave and gag. They believed me! I had to tell them I was kidding before they raced to the bathroom to vomit. (My nurse friend says that new moms have brought her boobie milk cookies a few times. Eww. We are weird out here in California.)

By then, Paul and I had settled into a comfortable routine at home, and thanks to his flexibility and our family and friends, childcare smoothed out. The love I felt for Stella started to expand and spill into every aspect of life, and made everything profoundly more joyful.

I was so relieved not to feel like hell, I doubled down at work. Everyone expected me to be more careful and reserved as a mom, but

I didn't become more cautious *at all*. If anything, I was more willing to take risks because I wanted my daughter to look at me and say, "You are a badass, Mom." I set out to tailor my life to be the woman I wanted my daughter to see and become.

I would feel sad (normal sad) to be apart from her for long stretches, so I started bringing her to work with me. When I had to be on camera, I'd set her up with a crew member, put some cartoons on the laptop, and tuck her into the "Costanza," a mattress under my desk for secret naps. Then I'd return to hang out with her for a little while, go back out and weld something. Everyone loved having her around. She was like a *MythBusters* mascot. And once again, all was happy and right with the world . . . until the next time.

IRONICALLY, the happiest time of my life, bringing Stella into the world, was also the saddest.

What can I say? I tend to go to extremes. When I became a mother, I learned what it meant to push past my limits, to sacrifice and do whatever was necessary for the greater good at work and at home. I fought back against the worst misery I'd ever known, and gained strength by just fucking doing it. You don't know how strong you can be until you have to be.

TOUGH TIMES DON'T LAST BUT TOUGH PEOPLE DO

CONCLUSION

If you have a physical ailment, it brings sympathy, but mental dysfunction makes people uncomfortable. When I was younger, I was so ashamed to think of depression as a mental illness that I grasped around for reasons to explain why it was happening, and blamed someone or something for it, no matter how small, like a school hallway diss, a bad grade, a random criticism.

Now I know that depression is a chemical, physical problem in my brain that I am powerless to control. It's like the mind control test experiment, when my brain signals were wired into Tory's arms. I laughed when I made his arms jerk around, spilling wine and meatballs on himself. It's not so funny, though, when your own mind has a mind of its own, and sends the signal to feel miserable.

By writing this chapter, I'm taking a cue from millennials who were raised to believe that mental disorders like anxiety, depression, and OCD are problems in their brains, and not "all in their heads." They have learned to grab the tools at hand and talk to their parents and others about what's going on.

I've hidden my depression for long enough. The final step for me in managing mine is to admit to loved ones and the whole world that, a few times over the course of the year, I will feel depressed, and that's just part of my life. I hope my friends, family, and fans will understand that if I need a little room, it's not about them. And if I need to talk honestly about my feelings, they'll listen and try to understand. For all I know, they've been suffering in silence, too.

If you are hiding, reach out to someone you trust. Therapy is expensive, but friends are free.

I found this wisdom in a fortune cookie. No shit. I cracked it open, read it, and thought, *This is actually brilliant.*

Tough times don't last but tough people do.

ONE MORE THING

Shout out to the neuroscientists who created the tech we used in the mind control episode. We had some fun taking their research and applying it comedically, but the reason Backyard Brains exists is to bring "neuroscience to the 99%," as they like to say. There are so many diseases and disorders associated with mental health that elude us. We are plagued with everything from autism to Alzheimer's. By making neuroscience accessible and getting people interested in the field, Backyard Brains is doing what *MythBusters* did. Amateur astronomers have discovered stars. Amateur paleontologists have found dinosaurs. Maybe an amateur neuroscientist will find the answers we need. In the least, maybe they will inspire someone to love the science. Who knows? Maybe a kid watching *White Rabbit Project* or playing with Backyard Brains kits will decide to become the neuroscientist that unlocks Alzheimer's or depression.

CHAPTER NINE

CREATIVITY

 hen I do lectures and events at colleges, the audi-
ence is full of eager young women, and the sight
warms my heart. I think, *Yes! The STEAMers are
here!* At the end of the talk, I always do a Q&A
session. The questions can be a bit predictable:

"What's your favorite myth?"

"Which myth was the most dangerous?"

"Are you guys friends in real life?"

But sometimes, a student asks what I call "a creative question," along
the lines of: "You know _____ [fill in the blank invention]? How did
you come up with that?" I love it when someone asks me about art and
creating, because, as I've made clear already, I'm not a scientist! I'm an
artist, just a curious person who likes to make stuff. Tory, Grant, and
I are all "creatives." Our job was to foster it at the most basic level—
start with nothing, and turn it into something. Every "solve" we rigged
began with the most elemental of all raw material: an idea.

People are naturally intrigued by the application of an idea, but
they're confounded by the generation of it. So I don't take the question "Where do you get your ideas?" lightly.

One myth I'll bust right now is that you are either born creative or not. Every single person is born to be a maker, a generator of ideas. It's true that some people might be gifted with natural talent in the arts. Talent is great to have, but it's not essential to creativity. All you need is a curious mind, and the drive to follow it wherever it leads you.

WHERE DO YOU GET YOUR IDEAS?

QUESTION

Distraction vs. Concentration

When I moved to San Francisco for college, the tech boom made the cost of living skyrocket. As a working student on a tight budget, I needed roommates. I found the perfect person in Bobby, a drummer, who was always on the road. He needed a place to live when in town and I needed to pay less rent. We became friends thanks to our similar tastes: a love for Jane's Addiction and Led Zeppelin, a decorating style that included skulls and candles, and the propensity to wear black or metallic clothing (the '90s).

Our apartment was just off Haight Street in the famous music- and art-fueled Haight Ashbury district. It was a one-bedroom that we could barely afford. I set up in the living room and gave Bobby the bedroom.

I remember my first night, sitting among my boxes, looking out the second floor window with all the lights off, and watching the street below, alive with voices and excitement. I was on my own, ex-cited, happy, scared, and home. I could be anyone I wanted to be.

I worked at a coffee shop near the cable car turnaround at Union Square. I had to be there for the morning rush, which meant waking before dawn. I would see all the same ladies of the night clocking out as I was clocking in. After charging up with caffeine during my shift,

I headed to school to get in as many classes as possible before going to my second job at a retail store on Haight Street called Backseat Betty. Their slogan: "Good things for bad girls and bad things for good girls," aka high fashion for strippers and rubber dresses for house-wives. I loved it.

Clearly, I was super busy with two jobs and a college course load, and appreciated my few hours of quiet in the apartment. When my roomie was home from his tour, however, he brought his band to live with us. Suddenly, I had no quiet time at all. It was pretty exciting to come home to a jam session but it was usually in the living room, where I slept. I would wake up in the morning, tiptoe over a sleeping guy or two, get dressed in my closet, and head off on my crazy schedule.

HYPOTHESIS

FOR REALLY GREAT IDEAS, YOU HAVE TO BE ABLE TO CONCENTRATE, ALONE, IN A QUIET ROOM, WITH NO DISTRACTIONS

I went from cherishing the quiet to embracing the chaos. The truth is, the guys weren't *always* around, and I had full power to kick them out when I needed to write a paper. Besides, I often got backstage access to great shows and occasionally came home to find a famous, attractive musician in my kitchen. Since Bobby was also a very good-looking guy, I made friends with a string of model-type gorgeous girls coming through the revolving door. They were always nice, and rarely ate my food. The apartment was like my own private Studio 54.

You would think my grades would have suffered, but I found that the more chaotic life was, the more I got done. My papers were more astute when I wrote them under the pressure of a time crunch. Be-sides, when I could get my work done quickly, there was more time to enjoy those living room concerts. I could draw pictures while they banged on trash can drums. We fed creatively off each other. Some of my best work came from a room full of music. Even amid the chaos, I was able to prioritize and multitask, a skill that I use every day.

(On the show, we once did a "battle of the sexes" test of multitasking. First, we asked male and female subjects to run through a series of chores in under five minutes, and then Grant and I tried the test ourselves. We had a healthy sibling rivalry and took any chance we could to battle each other. Grant was sure he would beat me on this one. He is insanely efficient—you should see his über-organized toolbox! He ran around like crazy to finish the chores in over four minutes and was so damn proud of himself. Then it was my turn. I calmly, coolly did it all—better than he did—in under three. He and Tory were shocked. I crushed Grant's time without breaking a sweat. #honestbrag.)

When you have no choice, you rise to the creative occasion.

My brain functions best when I'm working on five projects at a time. In fact, if I get stuck on project A, the only way to get unstuck is to pick up with projects B and C, and then out of nowhere, I get bright ideas for project A. When one part of my brain takes a break, another part gets busy, and that little break is all I need to spark something. This is just my process. It might or might not work for you. Figuring out how to focus and trigger bright ideas is a journey in and of itself. **If you go on an actual journey, you'll acquire previously inaccessible creative powers.**

(RITE OF) PASSAGE TO INDIA

CRASH TEST

When I was in India looking for grand insight (yes, I was *that* girl, with *that* hypothesis, that answers to every question could be found in far-flung exotic locales), to get to our destination of hopeful enlightenment, Dawn and I took a thirteen-hour bus ride with our huge backpacks on our laps. I was squished up against a broken window and sick the whole time from dense cigarette smoke, no personal space, and pins and needles in my legs from lack of movement. But hey, it was the cheapest way to get to where we were going and it was a true Indian experience.

We stayed in an ashram in Rishikesh near a famous hanging bridge called Lakshman Jhula, and drank in the exotic, beautiful place, the orange and rosy colors of the light from dust in the air and the warm aesthetic of the people in their colorful clothes. Every walk down the dirt road was a sensual experience, sights, sounds, smells, good and/or putrid. You would take a deep perfumed inhale of Nag Champa and then be choked by rotting garbage. The ashram had a sweeping view of the river from a small balcony. At sunset, a nearby temple rang bells and called to prayer from loudspeakers, filling the air with a reminder that we were a long way from home.

I loved the day-to-day ritual of our new life, in particular, the super-sweet and dense Indian coffee served in tiny cups that I drank on the way to our morning yoga class. It was a precious moment and feeling every day, and an important part of my Indian experience. I cherished the nightly ritual of little baskets set afloat down the river like lanterns, and became familiar with the people I would see along the road.

Our yoga instructor was a man named Rama Krishna who, along with teaching asanas, would massage your disrupted auras. He had a long white beard and a smile that was always filled with the beginning of a chuckle. We hung out with him and became friends. Dawn and I were in search of a true meditation experience and hoped he could help us find a Vipassana teacher. That is a completely silent meditation (all of my friends reading this right now, seriously stop laughing). After a couple of weeks, he asked us, "Why do all you Americans come to India to try to be Indians? You are American, you cannot find your meditation in India, you have to find your meditation in America." He had a way of explaining things with a charming smile that made you listen.

181

"And you," he said to me, "you are always painting. You are an artist. That is your meditation. You will never be able to be quiet. You must paint."

I thought, *You know what? He's right. I'm gonna just stay here and paint.*

The day after Rama Krishna set me straight, I walked to the stand where I drank my special, exotic Indian coffee. I got up extra early, excited for my new mission to find myself as an artist. Because I got to the café so early, I saw the coffee being made. This giant hairy, sweaty man, without pants on, stirred ingredients into a large vat. I saw him pour them in. My morning enlightenment in a cup was Nescafé and condensed milk!

All of my notions about the spiritual exoticism of yoga, meditation, and magic coffee were shattered within twenty-four hours. I cried from my third eye about it. Kidding. I hung around that village for another two weeks, painting sunsets and taking photo portraits of people on the street—and I did find myself as an artist. India had nothing to do with it. When I got home to California, I kept the creative vibes flowing by turning all my travel photos into sculptures and other cool things.

My quest for enlightenment seems kind of ridiculous to me in hindsight. Now I know that you get "answers" (practical or mystical) by asking questions and experimenting. For me, yoga isn't about aura massaging and chanting words I don't understand. It's the solution to the problem of staying fit at forty, building strength, and quieting my mind. I still paint, but my meditation is bigger than that. It's about meeting the challenge of being myself. **The creativity journey is always inward.**

When I set out on this yearlong trip, I knew I was in for an adventure and hoped to come back changed, but I couldn't have guessed how. I thought I'd find deep answers to metaphysical questions. Didn't happen. I did come back brave, with newfound confidence, and freedom from self-judgment.

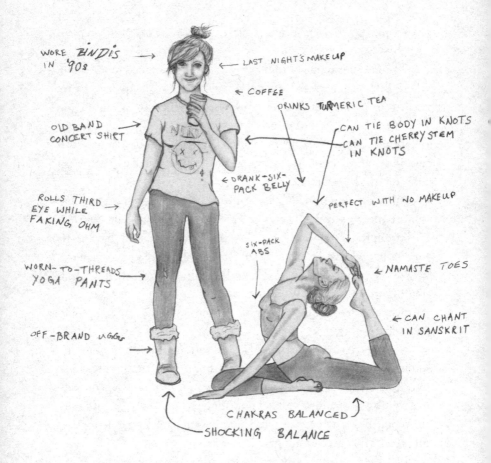

WORE *BINDIS* IN '90s

LAST NIGHT'S MAKEUP

← COFFEE

DRINKS TURMERIC TEA

OLD BAND CONCERT SHIRT

CAN TIE BODY IN KNOTS
CAN TIE CHERRY STEM IN KNOTS

← DRANK-SIX-PACK BELLY

PERFECT WITH NO MAKEUP

ROLLS THIRD EYE WHILE FAKING OHM

SIX-PACK ABS

← NAMASTE TOES

WORN-TO-THREADS YOGA PANTS

← CAN CHANT IN SANSKRIT

OFF-BRAND UGGs

CHAKRAS BALANCED

SHOCKING BALANCE

PS. Incense is carcinogenic. #justsaying

Live the Dream

For twenty-three years, I was a shy, artsy girl from Los Gatos, creative but unfocused, full of self-doubt. And then, while traveling, under the influence of beautiful new sunsets, grand adventure, and a new best friend in Lisa, I was inspired to draw at a much higher level. I wanted to impress Lisa, and wound up impressing myself. I stopped worrying about how good I was, and finally turned into an artist.

I got home to California a changed woman. I'd grown as an artist and a human being. My spirit was free, and my wallet was empty. Fortunately, if you were going to be young, artsy, and broke, San Francisco was the best place to be in the entire world. My equally poor and creative friends and I aspired to live as cheaply as possible. That meant back to the side hustles. We made jeweled barrettes and accessories to sell at local boutiques. I turned thrift store men's neckties into ornate necklaces.

I still had the travel bug, so I would save every penny and run off on small American adventures, bringing my side hustles with me. For example, I desperately wanted to see the Jazz Fest in New Orleans, but only had enough for a plane ticket. Solution? I painted postcards and sold them to tourists so we could buy beer, crawfish, and tickets to the festival. I provided an experience, engaging buyers in conversation about art and music, and gave them a moment of connection with a friendly, spunky girl in the red dress who made them laugh. I was part of their fun memories of their trip.

It sounds corny, but it was the first time I was an actual working artist. It wasn't a lot, but I was getting paid for art. Full disclosure, the postcards weren't that good but most people headed to music festivals are a bit off their heads. **It is possible to survive on art alone.**

Creativity was my occupation, as in, it kept me occupied during some long months of unemployment, when I was hoping to find a decent job with a big enough salary so I could go out to dinner once in a while. If you want creativity to be an occupation in the sense that it is your profession, that you are a member of the "creative economy," then proceed to bust your ass and make huge sacrifices to get your foot halfway in the door.

BE CREATIVE ON COMMAND

EXPERIMENT

For the last fifteen years, my job has been invention. Every single myth tested required our creating a prototype, because we couldn't go to a store to buy a sword-swinging robot or a face-slapping machine with custom remote triggers. Once, Tory, Grant, and I were tasked with making a grizzly gizmo featured in a kung fu movie called *The Flying Guillotine*, a weapon that could be thrown through the air, Frisbee-like, fall down around the victim's neck, chop off his head, and then capture it in a sack or net to be retrieved.

Needless to say, all three of us, though full of geeky glee, were stumped how to make a prototype for a deadly decapitation weapon, since no blueprints or downloadable instructions existed. We said to each other, "How the hell are we going to do *this*?!?" Only one way: trial and error, emphasis on error. We decided to work separately on our own designs and see whose worked best.

I started by going through the shop's aisles of shelves with random bits and pieces (some might describe the boxes of string, pipes, machine parts, and plastic bits as junk, but one man's trash is another's brilliant idea waiting to happen), and pulled out a fan from an air-conditioning duct. It was too heavy and complicated to work. I tried making a spring-loaded weapon, but the snap of the springs was scary. I might lose a finger trying to chop off a dummy's head. Not worth it! I eventually went with a head-size cigar cutter concept with a single blade. I put a giant, super-sharp kitchen knife on a couple of sliders that would decapitate the fake head with a good yank of the attached chain, and then capture the head in a net that would unfurl from a beautifully decorated dragon hat.

It didn't work. I was able to throw it over a dummy's head (after way more effort than the edit presented), but when I tried the guillotine to cut through the neck, it didn't slice cleanly. Tory's saw-inspired Frisbee of Death came close enough to working to declare the myth "plausible." I was satisfied with that, horrifying as it was. The process of problem-solving to get that far using only materials we had lying around the shop was a grim, gruesome, cool accomplishment—and a super-fun shoot.

CRACK UNDER CREATIVE PRESSURE

CRASH TEST

You know the story by now. I begged my way into an unpaid internship at M5, and, by chance, got the job as the host of a science entertainment show. The job itself was to create original, inventive, unpredictable rigs and gizmos every week.

And that's when I realized that it was one thing to paint love letters on a piece of sea glass for my boyfriend, and another to be tasked with coming up with a prototype of, say, a flying guillotine, in an hour.

We appropriated the model maker concept of "kit bashing" to create our prototypes. This is one of my favorite tricks I learned from working in Jamie's shop. It's the process of taking apart store-bought models and kits, or using random objects to create a new custom project. You can pull apart a model train and some plumbing parts, taking copper cables and a tire tread, and "kit bash" them together to build a robot. As an artist, I thought kit bashing was the perfect expression of creativity, and doing it alongside a bunch of guys who worked in the industry for so long was like a dream come to life.

INSPIRATION

Even though self-knowledge will take you to higher levels of creativity, you still need *something* to chew on to get the ideas flowing.

It could be **travel.** Just changing the scenery makes your brain think in different ways.

It could be a **movie, documentary, or TV show.** I was inspired by a series on the History Channel about serial killers, and made a collection of tiny dolls with realistic sculpted heads based on just their mug shots; the idea was to take all that big scary darkness and quarantine it in a wee small object.

I love going to **museums** and am a member at all the San Francisco biggies. When I feel a need to be inspired, my go-to is the ancient oceanic art exhibit at the de Young Museum in Golden Gate Park.

I trick myself into being inspired by **alternating between two mediums**—sculpture and black powder art. What is black powder art? It's a controlled explosion of black powder, a small amount of the same stuff that makes bullets fire, that leaves a shadow or impression on wood or paper that I look at, meditate on, and allow an image to surface in my mind that I fill in with paint. I apply myself to what's there. It's a good way to push through my mental creative blocks I might have while making sculptures. And vice versa.

Walking around always brings my eye to some new, cool thing that triggers a notion or idea. Some graffiti, a person on the street, a snatch of conversation, a random image, a piece of junk in someone's garbage that I drag home and turn into jewelry or a frame or something else. It's all out there, just waiting to be noticed.

Halloween, just the thrill of it, inspires me. October is Christmas around here. I design and sew all of our family's costumes and they can get pretty ornate. One year, I was a Death Queen, with a long white wig and skull headpiece. I was once a zombie with red contact lenses. Stella vetoed that one. "Too scary, Mommy," she said. I replied, "I can't wait until you're older," and switched to a fairy, which she approved. One year, we went as mother-and-daughter lemurs and Paul was our gorilla sidekick.

189

Big-eyed, I'd asked everyone I met for their war stories. "You worked on *Gremlins*? Tell me everything! How did they operate the robots? What inspired the look?" I was so eager to learn from all the old guard masters. I tried to remember everything they taught me and closely watched their techniques. It was an education you can't buy and it really helped in making prototypes for *MythBusters*. We'd look at our pile of broken machines and think, "We can take that spring from that, and the metal arm from this, attach a rubber hand, and voilà! We've got an automatic slapping machine!"

When we were shooting the show, it never felt like anyone besides the sound and cameramen would ever see it. We had such a skeleton crew that it was more like making home movies. After a few years, the slightly awkward TV practice of narrating out loud whatever you're doing became old hat. In fact, I'd hear myself doing it at home while making breakfast. "That pan is too hot. Never enough coffee around here. Whoa, that's how toast gets burned! . . ." My husband had to remind me that every thought that popped into my head didn't need to be narrated.

The point is, since I was so used to the crew and the practice of creating on camera, it felt like I was alone in my studio, or just playing at building with my friends. Being creative on command never felt like pressure. It was the most fun part of my job, and ideas continued to flow like a river.

But, when the show got popular, I started doing science fairs in front of an audience. With a crowd of people right in front of me, I found it stressful to invent on command. If I have to face creative deadlines (like writing this book), my anxiety kicks in, and it can be paralyzing. I'm used to comfortable creativity in my little studio in my backyard to make whatever moves me, or just sitting there for an hour, cleaning my brushes. But put me on a stage or under a deadline, my river of ideas might freeze over.

That does not stop me from doing it! I go for it, helmet first, and crash test through my anxiety and stress. If the result is lame, so be it.

190

When it's just for me, my art is better. But you can't always dance like no one's watching, or write like no one's reading (since Stella was born, I'd really like to just poop like no one's watching). People are watching and reading. So you do—and hope for—your best. **Any creative endeavor, no matter how successful it turns out to be, should be celebrated for the accomplishment it is.**

I have a tradition of going to a local ceramics place, and hand-painting mugs for every new career or creative venture that Paul or I begin. So I have mugs commemorating all my TV shows. We have one for his skateboard business, one for this book. The vessel is a good metaphor, too. We pour our hearts and souls into our projects, and coffee—there isn't one project I do that isn't entirely fueled by thick dark roast.

ANALYSIS

THERE WILL BE TIMES IN YOUR CREATIVE LIFE WHEN THE IDEAS AND INSPIRATION DRY UP

Break Through Blockage

I was in the midst of a long period of going in the studio to make something, and cleaning my brushes for hours a day, increasingly frustrated. I just didn't feel excited about anything. I was starting to get an antsy depression about not creating . . . and then, while watching Stella gnaw away on her fingernails, a terrible habit that she had to break, I thought about all the mom gossip about her classmates, the nose-pickers, back-talkers, the candy-sneakers. The light went on in my brain, and I latched onto my next big idea—a series of bad habit monster characters that children needed to conquer. Stella and I worked together on creating them. I did a workshop at her school and brought in some sculptures of the monsters. The kids went crazy, and I got that hit of having invented something that people could be excited about. Luckily, kids are disgusting, so I have a whole room of little bad manners monsters.

191

I look at Stella and get a contact high from her wild imagination. I'm not sure yet if she's inherited her parents' compulsion to create, but there are signs. Not too long ago, we both had colds and didn't want to take the dog out on a walk. Stella said, "Mom, let's get the stretchy string, tie a ball to the ceiling, and the dog will throw it for herself." It totally worked. I couldn't have been more proud of her.

On other days, I hope she won't be a starving artist. How amazing would it be if she could fix our computers or do our taxes one day? She announced recently that she wants to go to coding camp. She might be like the Alex P. Keaton in our household of liberal artsy types. How cool would that be? **If you are ever in need of a spark, watch a kid, or look at the world with the wonder of a child.**

Whenever I feel blocked, I look at Stella and am reminded of magic. If you don't have a kid in your house, borrow one from a friend or relative. Or, just zoom back in time to when you were a crafty kid who, like me, turned empty Cheerios boxes into 3D art, or wrote long, complicated dramas for your stuffed animals. Creativity just bursts out of kids, and ideas are like exploding firecrackers in their minds. I'm not saying you should steal ideas from a kid (although artists have certainly done it before), but it can't hurt to question everything, laugh hysterically at farts, and make weird connections, as if the magic of life hasn't been hammered out by adult logic.

CREATIVITY IS CONTAGIOUS

CONCLUSION

Any creative life has to begin with the one big idea, one that eclipses all the little ideas that people ask you about at college lectures. It's an idea you have about yourself and who you are (not what you do).

CREATIVITY IS CONTAGIOUS

When I was in India, Rama Krishna said, "You are an artist." He could say it a lot easier than I could admit it to myself. Saying "I am an artist" takes some conscious doing. A lot of people would roll their eyes at that statement. It can sound a bit pretentious. But since my whole journey is about learning to be brave, by calling myself an artist I am forced to live up to the title. I strive to be a translator of experience and emotion into art—or writing or prototyping—that people relate to, appreciate, and are moved by. You won't know until you declare yourself. Saying the words sets the intention.

As you progress in a creative life and/or profession, you'll find that one creative pursuit inspires and fuels another. The very act of making— be it dinner, art, a book, a pitch—gives me energy and ideas about other things to make. I love to visualize a mountaineering quote my friend's father used to say: "When one climbs, the rest are lifted." When life is full of creativity, it's vibrant and blooming. The best part is that it also spreads to those around you.

193

For that Flying Guillotine episode, we knew going in that our gadgets would probably fail, and yet, we did it anyway, not only for the job, but for the challenge and joy of creating. I see the same joy and passion in kids on the science fair circuit, and in the eyes of geeks everywhere. I hope to keep it alive in myself. My weird, wild life has let me stay a big kid. We are all born curious and fascinated with the world. In my mind, "growing up" always meant I had to stop having fun, but as long as I keep fostering my curiosity, I can be a grown-up *and* keep my giggles forever.

I could be happy taking a walk or watching TV every night, but I have this screaming need to create and paint. Creating is humanizing, it makes me a better person. I believe art is just as important as science, and that's why I push to add the A to STEM and turn it into STEAM. If not for novelist Mary Shelley, we wouldn't have the grandmother of science fiction. Ada Lovelace, Lord Byron's daughter, a mathematician, had the idea of using numbers to make pictures, and basically invented the foundation of digital computing. Use both sides of your brain. Foster art, combine it with science, and you'll get something that is truly groundbreaking. Innovators with artistic hearts are the ones who'll move us forward. Creativity is contagious (thanks, Einstein). Pass it on.

SETBACKS

The most controversial *MythBusters* episode ever became known as the "Cannonball Mishap."

The objective of the test was to see if a cannonball made of stone could be a viable weapon of carnage on a castle wall. We were on the bomb range and doing our control tests with a cannon we'd built. The day was pretty standard—for us, that is. We built our target, shot our cannon multiple times, collected results, goofed around taking silly selfies, just another location day.

The next thing we knew, a fire truck arrived at the bomb range. The fire chief got out and asked, "Are you firing cannonballs up here? One of them just tore through the neighborhood outside."

We thought he was joking. "You're kidding, right?" we asked.

"No, really. It went through the wall of someone's house."

And then we saw the news choppers.

We'd tested cannon fire there many times, and it was never a problem because we had so many fail-safe measures in place. First, the balls went through water barrels, then a brick wall covered in truck bed liner, then more water barrels, and finally, a backstop giant wall of dirt. Every cannonball we'd shot over many years always embedded itself into the giant dirt mountain that surrounded the range. On this particular day, we'd tripled up on safety by creating those extra layers to catch the ball.

What went wrong? In an almost inexplicable series of events, the cannonball bounced up the hill and launched into the air in a perfect ballistics trajectory. Then it arched down, into the adjacent neighborhood, crashing onto a sidewalk almost half a mile away, before bouncing up into the front door of a house, ping-ponging up the stairs through a bedroom and out the back wall, then bouncing off a roof in the next neighborhood, and finally coming to rest after crashing through the windshield of someone's (thankfully unoccupied!) minivan.

I didn't even know there were houses back there. When we first started using the bomb range, we climbed to the top of the hill and saw nothing but fields. So when they told us the cannonball hit houses, we were like, "*Where?*" We went up the hill and saw that, in the years since we'd first looked, development had moved closer and closer to the perimeters of the range. There were houses about half a mile away.

If we had come across this story, we would have assumed it was an urban legend. That kind of wild flight did not fit anything we knew about ballistics. Each impact, on the road, on the houses, should have absorbed the cannonball's energy and slowed down or stopped it, but it just kept going. We couldn't have replicated its path of destruction if we tried. There were too many variables. It would have been like firing a gun freehand through the same hole twice.

We found out right away no one was hurt, which was a huge relief. But I still felt the weight of the accident, and the implicit danger. It was *heavy*. I felt responsible. My whole body shivered and my breathing was shallow.

Right away, we shut down production. I sat down in the bunker with my hands on my head, listening to the news choppers circling above. The next day at the shop, we had an in-house therapy session and talked about what it all meant and what we had to do to make sure it never happened again. I just busted out crying (which I never do at work as a matter of principle). We could have seriously hurt someone. Even writing this, I feel tears building in my eyes.

I took it to heart, and many of my colleagues felt the same way.

We did a tour to personally apologize to anyone and everyone who would listen, and went into the exact community where it happened for a town hall meeting. They were, understandably, very angry and let us have it. I looked them in the eyes and let every verbal punch land square in the face. We hosted a barbecue and said, "We know it's not enough, but here's food?" Adam and I went to their local science fair and talked to the community, trying to convince them that we were so sorry and doing everything we could to prevent it in the future.

We never shot a cannon there again, only set off explosives. No trajectories.

We fixed their houses and the car immediately. The lawyers took over from there.

Strangely enough, the story went wide internationally. I had friends in every country from Australia to Russia to Mexico emailing me that they saw it.

But it didn't kill the show.

The cannonball incident did give us all a wake-up call. It radically changed our safety procedures. We had done dangerous things so often, maybe we lost some of our reverence for the implicit risk.

The lesson of the cannonball mishap: Don't get too comfortable.

QUESTION

HOW DO YOU BOUNCE BACK FROM SETBACKS?

No Crying in Science Entertainment TV

Early on, I was often worried that, one day, the producers or the network would wake up and realize I had no business being there, and they'd tell me to hit the road. I wasn't an actress and was mostly making this whole hosting thing up as I went along. My fatalistic view of my job meant I'd do virtually anything they asked.

REGRETS,
DEFINITELY
HAVE A FEW

Although the departure was rocky, I have no regrets at all about my tenure on *MythBusters*. I made best friends, had a tremendous amount of fun, and learned a million facts that I can spout at will. And even if it had gone balls-up earlier than it did, I would have been glad I went for it. I believe in learning from your mistakes and I have plenty to draw from. I tend to make grand life decisions very quickly, and find it easier to pick up the pieces later on rather than to prevent the awfulness with careful thought beforehand. I guess you could call the behavior "impulsive"? It's part and parcel of the crash test life-style. You have to own your actions *and* your emotional reactions to what you do, even if they haunt you still, years later.

Some of my regrets:

- Not buying a bigger house in San Francisco before the tech bubble.

- Not getting a job at Google in 1998.

- Not buying shares in Google in 2004.

- Ghosting sweet boys in my youth who didn't deserve to be treated so shabbily.

- Piercing my nipples. Big time.

- The memories I don't have because I blacked out from drinking.

- Every moment I wasted waiting for something to happen instead of being mindful in the moment.

- Swerving to avoid a fox in the road and totaling the best Toyota pickup truck ever. Fuck that fox!

- Pretending not to like it when Nicky kissed me in the sixth grade during spin the bottle.

- Giving away band T-shirts and weird clothes that could have been repurposed.

- Stopping my travel journals when I stopped traveling.

- Not writing more letters to my Aunt Betty.

- Eating fish tacos in Las Vegas.

- Sparkle blue eye shadow in the first few seasons of *MythBusters*.

- Attempting to rock a bindi and lace fingerless gloves in my midtwenties in San Francisco. Scratch that, I still have the gloves.

- Not learning to play guitar or to skateboard.

- Any time I didn't stand up for myself.

For example: I was a vegetarian, and yet I took on tasks with animal parts on a weekly basis. I once repurposed a pig spine, sewing skin to it as a stand-in for a human neck so that I could try to demonstrate decapitation with a sword. My gag reflex shivered with every stitch. I was spared from participating in the "Tastes like Chicken" challenge, when Tory and Grant had to eat gator and rattlesnake, but I had to both prep the samples and watch them eat. I did participate in a gross eating experiment about the myth that you get cold hands and feet when you're afraid—the theory was fear rushes your blood to your vital organs. To test it, we had to create a scary situation and then use a thermal camera to take our temperature. I sat down at a banquet of disgustingness: chicken feet, rotten intestines, and fish heads—scary foods, to be sure. I figured the lab-grade bugs were the cleanest and chose a cricket. I put it in my mouth, and made the mistake of not chewing before swallowing it right away. It grabbed onto the back of my tongue and top of my throat while struggling for dear life against my uvula. I wound up puking. There was a bucket under the table for just this outcome, how thoughtful. Since they'd rigged it with a camera at the bottom, I have to assume this was the outcome they'd hoped for.

Every time I puked, or almost did, it was caught on camera and I was known for the intensity of my reactions. I got less grossed out by meat over the years, but sometimes I played it up anyway. It became my schtick and it made the producers and crew so delighted to torture me.

I did manage to swallow a maggot, though. (Pro tip: If you have to swallow a bug, look for soft, legless creatures like mealworms.)

FYI: A production assistant in charge of cleaning up didn't know what to do with all the gross food from the cold feet banquet, so he shoved it in the workshop fridge. The power went out over the weekend and, on Monday, when someone opened the fridge door, the stench was so overwhelmingly putrid, everyone in the shop started dry heaving. We had to evacuate the entire place and the poor assistant had to clean out the fridge and scrub it down with bleach. Ah, the good old days of being underpaid for the most vile jobs. I felt for him, but I was sure glad I'd already paid my dues.

HYPOTHESIS

IF YOU KICK ASS AT WORK, REALLY GIVE IT YOUR ALL, YOU MIGHT HIT SOME BUMPS BUT YOUR JOB IS SAFE

I never took sick days because too many people relied on me to be there. I showed up, no matter what. I worked through sickness and injury, smiled through migraines and sniffles. Once, we were doing an episode about guns and needed a ton of ballistics gel blocks (a gelatin-based material that you make like a Jell-O mold). We had to mix it in five-gallon buckets and let them set overnight. We were in a rush, and I skipped lunch to get them finished. In my hurry, some spilled on the floor. I slipped in the warm puddle, fell down, and dislocated and cracked my knee on the floor (my bad knee). I screamed and rolled in pain, covered in gooey, gelatinous ooze.

TOP TEN WORST SMELLS IN GLAMOROUS TV

The rancid fridge was bad. These were worse:

1. Wolf urine
2. Goat horn on the sander or band saw
3. Large graduated cylinder of week-old saliva
4. Shark skin curing in urine
5. Rendering plant where we took bullet-ridden pigs
6. Blended cow brain
7. Shop tools with embedded decaying pig carcass
8. Vat of emulsified cow manure
9. Lion poop in jar
10. Human poop in a beaker

(Luckily, I was on maternity leave for the human earwax experiment, otherwise it would top the list.)

A crew guy called out, "Call an ambulance!"

I said, "No, just drive me to the hospital." I couldn't afford the ambulance ride! The look of horror on the faces of the Australian crew that came from a country of socialized medicine? Priceless. I was back later to finish the episode on crutches and a knee immobilizer.

This was my life. Some days, you ate a maggot. Some days, you slipped on ballistics gel and broke your knee. Most days, I couldn't believe how cool and strange it all was.

In the world of adventure TV, there's a phenomenon called "Kodak courage." It's when you agree to do things on camera and take big risks that you would never let happen otherwise. Like when we did a show about injecting an RFID (radio-frequency identification) chip in your arm and seeing if an MRI would rip it out, I said, "Let's do it!" and rolled up my sleeve. What if the myth were true? I didn't know that I'd need surgery to get it removed, so that chip stayed in my arm for years—and totally freaked out the foil hat–wearing conspiracy theorists.

We had to be vigilant with safety specialists because, as soon as the camera went on, they'd be more lenient with their recommendations—and they were the experts. Sometimes, I look at episodes and can't believe we kept filming. On an episode of *Thrill Factor* (a show I did for the Travel Channel), Tory and I rode Falcon's Fury, a theme park ride that takes you up 335 feet on a freestanding tower and drops you in free fall straight down, face first. The ride is so intense, they don't let you ride it two times in a row because they're afraid the effects might make you pass out. I didn't find that little fact out until we'd done it *four times in a row*, and my legs gave out below me. I asked, "Why did you let us do it?" and they said, "Because you are on TV!" I fake laughed to cover my concern, punched blood back into my legs, and strapped in for one more time, as if we were immune to the laws of physics and the normal constraints of the human body (*smiling, eye-rolling*).

On *White Rabbit Project*, for the episode about "Tech Before Its Time," I road tested (literally) an early form of in-car navigation in a

1933 Packard. The producers sourced a vintage mint condition car. For whatever reason, I'd never learned how to drive a stick shift, a dirty secret of mine. I had informed the producer but he was sure the car didn't have standard transmission. I was pretty sure he was wrong and after one Google search, my heart dropped. It wasn't just a stick, but an old-timey one, with no instruction manual to explain the gearing. We were on a tight schedule, and there was no going back on the Packard, so I had to crash test drive it by studying YouTube videos and taking a one-hour lesson with this little old man who had an ancient, beat-up Toyota—on the day before shooting!

Nervous as I was, I walked into the workshop to a vintage car collector polishing the hood. "How's your stick game, lady?" he asked. "You're driving a three-hundred-thousand-dollar car."

I actually did drive that car through downtown Los Angeles rush hour traffic, grinding a few gears along the way, in a complete panic, thinking, "This is your fucking job. Let's go!" I sat up straight and kept a smile on my face. Every time the collector grimaced, I told him my modern car shifted a bit differently but not to worry, "I got this." Thank goodness I didn't crash this crash test. **At any job, where someone is paying you money to work, coasting is death. You have to dazzle, daily. Slacking off is not an option.**

Being a good sport, playing hurt, not crying, never phoning it in, working through migraines, pain, and depression, I did it all. After about ten years, I started to feel confident that I was safe at *Myth-Busters*. But I wasn't. Safety on TV is an illusion at best.

In 2014, the show had been running for over a decade while the TV landscape changed around us. The network had new leadership and they wanted to bring in their own vision. *MythBusters* had survived eight different network heads over the years, but this new one came from TLC, which was heavily into *Honey Boo Boo*–style reality. Our numbers weren't as strong as they once were, and our budget was cut dramatically. The producers did everything they could to keep the show intact, but it just wasn't possible.

The build team was offered a deal to work only three weeks per year doing some mini myths but with an exclusivity clause, which meant that we couldn't work for anyone else during the contract time. I'd be living under a bridge, feeding my daughter cat food (remember—my biggest money fear of all time?) if I only worked fifteen days out of 365. We asked them to strike the exclusivity so we could stay a part of the show, but they said no. It felt like an offer we had to refuse so it would look like we quit. If they fired us, our super-loyal fan base would have been up in arms.

That was that. We couldn't agree to their terms and it was over.

It was such an unceremonious, perfunctory ending. Instead of coming back from Christmas vacation, our crew and staff were let go. Our shop was shut down and the locks were changed while we were away. I had to *schedule* a time to get my stuff. While I packed up my toolbox, a staffer followed me around to make sure I didn't steal anything. It was kind of like, "Don't let the door hit you on the ass on the way out." I'd grown up on *MythBusters*, and this was a sad, sorry end.

I felt disrespected and undervalued by the production company. They didn't realize how much we shaped the show, the way things were tested, how we came up with the myths. Even the logo was a shop design. I can't speak for Tory and Grant, but my feelings were hurt by how they dismissed us. Now I know that this kind of thing happens all the time. But I didn't know this side of showbiz back then. I'd never been fired from a TV job before. It felt like I'd been dumped and I took it personally.

Out of loyalty, Tory, Grant, and I publicly toed the line about the circumstances of our leaving for the sake of the show. Then a producer implied in a magazine interview that, in diva fashion, we'd demanded too much money and they had to cut us loose. A revisionist history! At the news of our leaving, the fans had been up in arms, starting petitions and taking to Reddit. I guess the producer felt attacked, and changed the narrative. I was annoyed as hell and commiserated with the guys. But what could we do? Get in a Twitter war over it? I let it go.

When the show got canceled soon after we left, the fans blamed it on our exit. Not sure that is true but it warmed my heart that they loved and appreciated us so much.

Tory, Grant, and I were asked to return for the final episode, and I found it extremely hard to come back, make nice, and pretend like they hadn't treated us shabbily. In the end, I didn't agree to appear in the finale for the network or the producers, or for myself. I did it to give closure to the fans, the millions who watched *MythBusters* and were supportive of us. If they hadn't been there, I wouldn't be here. Our show was very collaborative with fans that supplied us with myth ideas, sent letters that we read in recap shows, and engaged with us at conventions and promotional events. The final episode was my good-bye to them.

I cried during the filming—breaking my own rule for the second time! (more on that in the next chapter)—and felt the magnitude of what the show had meant to me, and everyone involved with it. It'd become so much bigger than any of us could have imagined and we'd made an impact that extended beyond the show itself. During the show's run, at college events and conventions, young women would come up to me, hug me, cry in my arms, and tell me that watching me on TV was the reason they were engineering or chemistry majors. I was so fortunate to have made an impact on other people. I was just having fun doing my job and didn't realize I was creating an echo in a canyon that eventually girls heard and acted on.

During low moments post-*MythBusters* unemployment, I thought of those women, dried my tears, and psyched myself up. I'd have to push myself to do something else, something just as cool and impactful. This couldn't be the last thing I contributed to the world. I had a responsibility to those young women, and to the very young girl at home (Stella was four then), to carry on, be a good example, and not mope. I would push on, because I wanted all of those fans to persevere. I'd crash test into the next job, whatever that might be, and show them, and my daughter, that taking calculated risks and being brave pays off.

EXTENDED
QUASI-EMPLOYMENT

EXPERIMENT

Our firing wasn't known immediately after it happened. We were asked not to talk about it until episodes without us started to air months later. We were under a gag order, essentially, which I honored.

But, since people in the position to hire me didn't know I was available, they weren't calling me. I missed a whole season of shooting pilots for the following year. The frustration of not being able to make calls and contact people about future employment was agonizing. It felt like I was never going to work again. I'd cultivated a set of skills that were only useful for a mythbuster, or maybe a secret agent. What was I going to do next?

When our ousting was finally announced, I got some phone calls asking about my availability. I was overcome with relief that people wanted to work with me. *I'm going to get another job*, I told myself.

I started setting up meetings with production companies and networks in my area—Nat Geo, History, Travel—to pitch ideas or just to say, "I'm here!" I talked to everyone I could, pimping my proverbial ride. I pitched the same idea, slightly adjusted for whatever network I was talking to, and I got shot down a lot. *You don't like that one? I got a million others! A computer full of concepts! Let's talk!* It's intimidating to go into a room of network execs and convince them to spend serious cash on your idea. But I'm not the type to rest on my laurels and wait for my future. I will write my own life . . . and maybe even a book (spoiler: you're reading it right now!). **Even if you get an "I love it!," it might still be a "We decided to go in a different direction."**

In a fit of unemployed madness when I hadn't gotten a solid offer right away, I thought, *You're going to have to get a real job*. That would require a grown-up wardrobe. I went straight to Theory and purchased black slacks for my future "normal" employment, whatever that might

EXPERIMENTS FOR
THE UNEMPLOYED

When *MythBusters* ended, I got very depressed. I needed to use my protocol not to sink into a long bout with it. What worked:

- **Mountain biking.** A physical adventure was a ready excuse not to think about work. I'd say to myself, "When the ride's over, I'll start stressing again." So I'd just keep riding across the Golden Gate Bridge, up Hawk Hill into the Headlands, and back home—some serious hill climbs along the way. It took hours and killed me, and yet, I live. The endorphins kicked in and nothing seemed so bad anymore. When my anxious mind went to sleep, my creative mind woke up.

- **Taking a mental health day.** Sometimes, I'd take a day to breathe. Life can mow you down, and the impulse is to get back up. But occasionally, it's better to stay down and regroup first. I'd been so strong, for so long, working long hours and taking care of my family. It was okay to give myself a day not to do anything but take a walk, have a long lunch, watch Netflix, draw with Stella. The next day, I'd be ready to do battle again.

- **Baking.** I dove into baking, and learned the chemistry of it. No one can touch my caramel corn. I made some crazy cakes, whipping the eggs separately and folding them in to make fluffier batter. I can decorate with fancy flowers and make cakes into sculptures, never using a bit of fondant. (IMHO, fondant is cheating, and it tastes horrible.)

- **Cooking.** I even learned to cook food I won't eat, like crispy bacon and a perfect roast chicken. I approach it all like a myth. I research all the "best" ways to cook on the web. I combine and try all the methods I find and test the product on my family and friends . . . and even neighbors.

- **Combing.** I would go to flea markets and collect odd things: old gears from broken watches, rusted keys, tiny roller chain. Then I cast them in resin for pendants or make them into earrings. I love to repurpose the beauty of *wabi-sabi* forgotten pieces and make them important, like resurrecting ghosts. My husband used to collect vintage photo albums full of lost memories and make them into paintings. I think this is a subconscious way of battling mortality on an artistic, esoteric level. So, yeah, I did a lot of deep thinking, too.

be. I brought them home, hung them in my closet, and expected to wear them at some office, in a cubicle or at a desk maybe, answering the phones, as my science/build dreams withered and died on the hanger.

My optimistic side thought, *Oh, please. Don't be so dramatic. People are going to start calling. Right . . . now.*

I spent a lot of time staring at my phone, willing it to ring.

Okay, I did make some pilots that sadly no one will ever see. I was hired for some one-shot specials, guest host gigs, and speaking engagements. It's not like I sat around binging TV. I called people for information interviews and asked them what was going on, and how they handled setbacks in their careers.

I just wanted to keep going, and the only way to do that was to put myself out there. I practically wrote my name on bathroom walls, saying, "Kari. Secret agent. Available for hire. Here's my number. Call me anytime!"

Anyone who can stay calm between jobs has my respect. Me? I go nuts. I've had one job or another since I was fifteen, and I didn't know what to do with myself without a place to be. I'm good at vacation, too, don't get me wrong. But this was different. I had a house and a family to support, and no serious offers were landing at my feet.

Tory and I teamed up and landed a series called *Thrill Factor* on the Travel Channel about amusement parks and the science of thrill rides. I love roller coasters, which might not come as much of a surprise since I live as if I were on one. (I actually got my period for the first time, in white culottes, on the Tidal Wave roller coaster at Great America during our eighth-grade field trip. I started wearing a lot more black immediately after. Damn centrifugal force and puberty hormones!)

Major downside: The pilot was about a waterslide, a super-steep waterslide. Water parks are a germophobe's vision of hell. There is not enough chlorine in the world to counteract all that urine. Maybe it's just me, but I don't like to swim around in places that not only are pee-filled swamps, but have to post, "Persons having currently active diarrhea, open sores, or communicable diseases are asked not to enter the water." I assume if they have to post the sign, there is history there. Just sayin'.

I risked my life for that show. I hope Tory appreciates that!

I assumed *Thrill Factor* would be an in-between-jobs job, a one-off pilot, that it wouldn't amount to much. But it turned out to be a series of ten fun-filled episodes. I learned a valuable lesson, that you have no idea what's going to happen, not only with the longevity of the job itself, but who you'll meet, the skills you'll develop, and what you'll learn about yourself. I learned, for example, that waterslides are, basically, a power enema using hundreds of people's pee. I also learned that Tory and I worked incredibly well together outside the context of *MythBusters*, and have been a dynamic duo on specials and series ever since.

THERE ARE NO GUARANTEES. IN EXPERIMENTS. IN JOBS. IN LIFE.

ANALYSIS

If you know that no job is one-hundred-percent safe, and that you will always have more to prove, you can bounce back and on to bigger and better things if (when) the current job ends. I had to learn that key fact first, before I could start moving on. My recovery was in slow motion after *MythBusters*. It was two years before I felt rock solid about my prospects again.

I covered projectile gourds on *Punkin Chunkin* for years, and *Large, Dangerous Rocket Ships*, as I've mentioned. I also MC'd robot competitions for kids, an annual highlight for me. Each kid's home-made robot was assigned a task, like doing an obstacle course or picking up objects. I loved doing these kinds of events because the kids didn't care who I was. They were just excited about their robots and wanted to tell you about them, how they came up with the design, how they put it on the wheels. The most beautiful thing in the world is a hyperactive and focused kid telling you about their creation.

For two years, I went from gig to gig, and managed to pay our bills, but it nagged at me that I didn't have something solid and long-term.

Then, out of the blue, the former production company of *MythBusters* was approached by Netflix to develop a new show that would get the build team back together, and *White Rabbit Project* was born. I was really excited to work with Netflix, the new frontier of TV. The concept of the show was taking a cool topic like famous heists or tech from movies you'd like to see in real life, and then doing some online research on it. Anyone who's done this knows what it means to fall down an internet rabbit hole. Hence, the title "White Rabbit Project" of taking those searches as far as we could, with adaptations and reenactments. I'm proud of the work we did, and we made ten episodes of great TV. But, sadly, there will be only one season. **The gig economy is the wave of the future. For me, the future is now.**

Onward. To what? I couldn't tell you. But I was, am, open to any and all ideas and opportunities. I started saying "yes" to everything. Every meeting, every part-time job, and sooner than later, I found myself working steadily at a bunch of different things.

It's All About Who You Know

Now that I'm going from gig to gig, I realize that having the same job for over a decade is unusual. Most of my friends are writers and artists, people who bounce from project to project. And then there are all the people in my city, and the world, who are starting their own businesses, and don't rely on the traditional model of being hired by a big company with all kinds of bonuses and benefits attached.

As members of the gig economy, we have to reinvent ourselves constantly, and network like it's going out of business—because if you don't, you might go out of business. I go to every panel discussion, charity event, and convention because I might meet interesting people at the Google Science Fair or the robot competition who could help me down the road or connect me with someone who might need me immediately. A phone call or meeting you have in May might bear fruit in September. A job that might seem like a placeholder could open doors that take you exactly where you need to go. A few times, I've done a random spot on a show, and met a producer who called me years later with an offer. I've landed more jobs by showing up at random events and talking to people than I have by sending out résumés and pitching.

Women need to learn how to network. It's not just who you know, it's who you'll meet when you show up, introduce yourself, and just chat. Don't laser target execs or high-up types. Talk to *everyone* who has something worthwhile to discuss. A receptionist today might be a CEO tomorrow. A production assistant today might be a network exec tomorrow. You just don't know.

I see half of my job as networking about future gigs. Last year, I was asked to appear at a college fair, and met the person, totally randomly, who was an author, and he suggested I talk to his agent to write a book about women being braver . . . and now you are currently reading it. It hadn't even crossed my mind to write a book, but I realized, as I talked to the agent, that I did have a lot of things to say about how to make it as a woman, a DIYer, and a crash tester.

This book, its very existence, is an example of what can come from randomly asking people how they wound up where they are now. The big mistake people make when networking is to ask for advice or to pitch yourself. People are far more interested in you if you're interested in them, not necessarily what they can do for you. A lot of women in power positions would love to tell you their story, because it's most likely a damn good one.

In general, while networking, don't ask, "What should I do?" to get a job, while between jobs, when figuring out what to do next.

Ask, "What did *you* do?"

Who you know might help you, but the real gold is in learning what they know.

It'd be nice to have a road map for success, and I wish I could point you in exactly the right direction. But the road map doesn't exist. As you plot your individual, totally unique course, it does help to talk to people about trail blazing. Their stories might inspire you to climb to the top of that wall, get that job, create that website, write that book. Your course to get from point A to point B might not be linear, so cast a wide net, be all over the map, and expect your destination to change.

If there's one thing I learned on *MythBusters*, it's the value of critical thinking. It's too easy to pigeonhole yourself by thinking you're one thing. I never would have thought of myself as an "author," but thanks to a chance encounter, now I am. I didn't think I'd host a travel-and-tech futurist show on Nat Geo, but I do. If it ends, it ends. But no matter what, I'm going to live and work on my own terms. I'm not the same plucky redheaded girl from *MythBusters* anymore. I'm a woman with all kinds of skills.

And FYI: The Theory slacks are still hanging in my closet, unworn. I told myself, "The day I cut the tags off is the day I stop trying." The pants are now a symbol of resilience and a reminder to stay flexible and be open. One day, I won't need the reminder, and then I'll burn them . . . or blow them up. Yeah, that's more my style.

KEEP WIPING

CONCLUSION

When you do anything in life, be it starting a new job, testing a myth, or shooting a cannonball, *failure is always an option*. You can hedge your bets by working hard, playing hurt, and being a valued team member. But you might still wind up on your ass, for a long time.

Failure isn't fun, but it does make you a better person. You grow a thick skin, and lose a lot of ego. When you're not weighed down by ego—a feeling like you deserve success—you can move faster, get stronger, and be more flexible.

Ideally, if you fail—on the job, getting a new job—you'll learn from it. If I do this, or talk to this person, what will happen? You learn what will and won't fly, and what won't fly now but might fly later.

When I was young, I struggled and took risks, had adventures, and met strange and wonderful people. I want to be that person— now and always. I'm happy in my beach 'hood townhouse; I have no desire to return to my crappy apartment. But risk-taking and adventures? Constantly meeting new people? Yes, please. I'll continue on with that dream. Every career setback has brought a wave of new ideas and people into my life, and I've grown because of it. My earnings have grown, too.

Here's the truth that women don't always want to admit: Safety at work can mean being stuck there. You won't double your salary at the same job, even with a promotion. To make giant leaps, you have to break out of the safe space. The safe space, I'm sorry to say, is a trap. Even *MythBusters* was a kind of trap. Now that I've worked on other shows, I've seen how it kept me in a box. I'm out of the box in a big way now, traveling the world and working toward being the boss.

Life is always going to be hard. When it stops being hard, it won't be interesting. Even if steady work would give me a sense of security, I want it to stay challenging, and have an interesting life. That's the larger goal.

If I've learned one thing, it's that shit happens. When I was a brand-new mom, I took Stella to MoMA to expose her to culture (she was an infant and had no idea what was going on; like I said, *new mom*). At one point, I caught a foul whiff and looked down. She'd made an explosive poop that went up her back into her hair. It was an unholy mess, and I wasn't prepared with a new diaper or a change of clothes (again, *new mom*). I ran with my shit-covered baby into the bathroom, totally panicking. I got a paper towel and started wiping. And another one, and another. I told myself to just keep wiping, and she'd be clean. So I kept wiping until Stella was a naked pink sweet-smelling baby. I put her back in the stroller completely naked, and walked out of there. The security guard saw her, and just laughed. Must have been a parent, too. Since that day, whenever I get in a pile of shit and have no idea how to get out of it, I think:

BRAVERY

The smell of fear—that people and animals can uncon-
sciously detect stress and fright in humans and know
to attack—is a cliché of pulp fiction. Of course, like
most movie motifs that sound too dramatic to be true,
there tends to be a basis in scientific fact. It was perfect
fodder for *MythBusters*. The first step would be to get some smelly
samples from our own bodies, so we put pads in our armpits to absorb
fear-induced sweat. And how did we induce fear? I got in a glass coffin
for seven minutes and was then covered with forty scorpions.

I was mainly glad it wasn't rats.

The glass coffin was so Sleeping Beauty. I just hoped I'd wake up
if one of the scorpions stung me. I was given safety goggles so they
wouldn't crawl in my eyes. Some comfort until they put one directly
on my face so I could stare directly at a barbed tail. I was just covered
with them, from head to toe. The handler was a pro who has done
snakes and bug wrangling for movies, including porn (snakes).

He said, "Just don't move and they won't sting you."

But I couldn't help it. I was shaking. My foot convulsed with fear
involuntarily. The scorpions started biting my nails because my orange polish was the same color as their food. I was mumbling in terror
because I was afraid to open my mouth or they'd crawl inside and
burst out of my stomach *Alien* style.

I tried to breathe shallowly and go to a Zen place, but I couldn't calm down. I made it through the seven minutes—no idea how—of sheer hell. Paul came to the set that day. He hardly ever did, but he had to see this. I think he was just as scared as I was to see me like that.

When I took the sweat pads out from under my arms, they were *drenched*, literally dripping with fear.

As the handler collected his bugs, he said, "I've never seen someone film with scorpions before. I was wondering how this was going to turn out. You're brave. I can't believe you stayed in the box for so long! Especially when they started to bite."

If he'd told me that ahead of time, I would have said, "Fuck this! I'm not doing it!"

LEAVE EVERYTHING
ON THE STAGE

CRASH TEST

When I turned sixteen, I was finally starting to feel like a woman, and so I wanted to wear a sexy costume for Halloween instead of the usual goofball options. My mom studied belly dancing when I was a toddler, and she had a lot of authentic *I Dream of Jeannie* colorful silky harem pants, and coin bras. I told my friends, "I'm going to wear one of those outfits to school!"

I had to screw up my bravery because my stomach was exposed and the purple, balloony pants were low waisted. I couldn't hide in them, even with a silk veil. I did my hair and my makeup and thought I looked pretty good. The only problem was that I wasn't filling out the coin bra. My mom was a beautiful woman with a more generous body than my budding sixteen-year-old figure.

I said to myself, "You have a problem in need of a solution." The obvious solution was to stuff the bra. I made boob inserts from fabric scraps I sewed into little balls, and they did make it look like I was filling out the bra.

So I arrived at school, belly exposed, finger cymbals clinking on my hands, and passed as a sexy dancer all morning. My school looked like Rydell High from *Grease*, with a big front lawn and a wide stairway leading to the Hellenic pillars at the entrance. The tradition was for the whole school to sit on the lawn out front during lunch to watch a costume parade/contest of kids walking down the stairs and striking a pose as an MC commented into a microphone. It was a "spirit day" kind of thing. I'd never participated before, but I felt good about myself and my costume, and said, "I'm doing it."

When it was my turn, I started walking across the stage. I threw my arms up in victory, and one of my bright white fabric falsies popped out of my bra, bounced all the way down the steps, and came to rest at the feet of my English teacher Mr. Hood.

I had to decide in a split second if I should go over and pick up my boob, or just leave it there. In that moment of acute mortification—refuting the myth that a teenager could actually die of embarrassment—I just left the stuffing where it landed, rushed off to the girls' room, changed out of my costume, and put on the regular clothes I'd brought just in case.

The period after lunch, I had English class, of course. I walked into the room with a face still as red as a hydrant from embarrassment. Mr. Hood came into the classroom. He was usually a quiet man but today, he had a big beardy grin on his face, the kind you couldn't suppress if you tried.

He looked around the class and said, "Sooooo, does anyone have anything to share? Kari?"

A bunch of kids burst out laughing. I thought, *I've got two choices here. I could: 1) start to cry, escape to the nurse, pretend that I have malaria and go home, or 2) act like it never happened.* I went with choice two. I sucked it up and acted like nothing happened. I blinked at the teacher and said, "I got nothing." He grinned and then got on with the class. It may have planted a seed for a future in television. After all, as we liked to say on set, "There is no dignity in TV."

I learned about option number three for dealing with potential embarrassment—laugh your ass off—while traveling after college. I was sailing down the Nile with Lisa, and she really had to pee. We got off our little boat, went to this sandy area, and she pulled her pants down. The dusty patch turned out to be a high-traffic camel and horse crossing. Four Bedouins rode by, saw us, and started laughing hysterically. I guess I should have covered up my friend and helped her, but we both fell over laughing, too. She got covered in her own pee, which only made the hysterics worse. Being embarrassed can be just as funny for you as it is for the people who see you in a compromising position.

Life is short; surrender your dignity. After you've been embarrassed a hundred or a thousand times, you understand that none of it matters, so you might as well enjoy every chance you get to laugh.

WHAT DO YOU HAVE TO BE AFRAID OF?

QUESTION

The Mountain

This story isn't about the perseverance of hiking *up* the mountain. That would be too basic. My story has a Byronesque twist: It's about climbing *down*. My mountain wasn't figurative. It was Annapurna in the Himalayas. You have heard of Everest? Annapurna is her neighbor, just shy of three thousand feet at her tallest peak.

Dawn and I had hired a guide to get us up and down the tenth highest mountain in the world in fourteen days. We made the long, beautiful, arduous hike that led us low into valleys with falling pink petals and green grass, and then so high we were cutting ice steps to get to the top. We had to carry everything for all weather on our backs. Halfway through, my feet were a mess—bloody, swollen, and sore, losing a toenail. My blisters had blisters. But we made it, and the top was fantastic. We had climbed so high that I had slight altitude

PUT YOUR SHIT OUT THERE

Back in my twenties, I thought I was smart, tough, and honest. The truth was, I was insecure and kind of an asshole. I thought I knew it all, but I knew *nothing*. Yesterday, I worked with a twenty-five-year-old girl who was so full of cockiness, the brand of arrogance that comes with being young, and I just looked at her, remembering my past self, thinking, *Oh, that's so cute. You have no idea that you know nothing.*

I did some dramatic writing back then, poetry and stories I emailed home to my mother, just to let her know what I was experiencing while abroad. She dutifully saved them. I pulled out a dusty box recently and read through my poems and journals. They are *beyond cheesy*! So many platitudes and clichés about "life" and "passion," as if I had all the answers, when, really, I didn't have a clue. I thought I was Jack Kerouac, writing *On the Road*. More like *On the Crapper*. I thought my poems were so romantic and deep, but, oh my God, they are embarrassing. Mom thought I sounded great, and was so proud, she sent them to the *Los Gatos Weekly*, where some were published (why, Mom, why?).

Reading those poems now is mortifying in a good way. Besides the hilarity of how bad they are, they provide truth and comfort. I'll probably look back at sixty on my forties and say, "You were *still* an asshole! You *still* didn't know anything!" and I'll love that, too. The promise of lifelong learning, even if it's just how big an asshole I am, is a gift that never gets old. I know most people prefer being young, but I *adore* being in my forties. I love the humility that comes with it. That's where I am now, and it's a really nice place to be.

sickness and didn't even care. Check that one off the bucket list. Now we had to get down and out before our permit expired or they'd charge you a fortune to come in and rescue you.

We had to rush, so we walked all downhill, no flats, on trails and stairs, overheating in shorts during the day and freezing in long pants in the evening. We stayed in a village on our second day of descent. I woke up at dawn, and got down off my cot, but I must've taken a weird step and my knee popped out of the socket. It was blindingly painful. I might've blacked out for a second. In my mind, I'd been telling myself, "You can do it!" But my body said, "I'm taking you out."

First, I had to get my knee back in the socket. I lay on the floor, grabbed my knee, and punched it back into place. I put a neoprene brace over my swollen knee while gritting my teeth so hard I thought they might shatter. I used some scarves to immobilize it as much as possible and fashioned a broken tree limb into a crutch.

We set off and every step was agony but we had no choice but to keep going. We were still high on the mountain, didn't have the money to pay for the rescue, and I was otherwise healthy. I'd set the goal of hiking Annapurna, and that meant down as well as up. No matter what, I was going to accomplish my goal. I might've been delirious by that point. I was definitely a bit outside my body, as if I were in a storybook. When I was just about to give up, two butterflies appeared in our path as if leading the way, and I said, "I'm taking that as a sign!" Butterflies as spirit guides? Why *not*? It was Nepal!

Another sign: We heard that a wild tiger had been spotted. I bet trekkers are *delicious*, so we decided to move faster. For days, I'd hop twice on the crutch for every one step on the "good" leg with the beat-up bloody foot. I remember stopping at cold water streams and sticking my knee and feet in the water to numb the pain.

I silently cried a lot and just when I started to give up completely— "I don't think I can go any farther!"—this horrible storm came in, with lightning and thunder, and I hobbled as fast as possible to the next village for shelter. Ahead of us on the trail, a middle-aged woman was walking along with ski poles for balance. I watched her get struck by lightning right in front of me. It was like she turned to stone and then fell over sideways, hitting the ground like a statue frozen in her pose. Her friends grabbed her now-limp body and carried her to the closest camp. She was sobbing all night long and every time lightning would hit the tin roof she would scream in terror and my heart would jump. I felt really horrible for her (but also kinda wanted her to shut up already). It was like being trapped in a tenth circle of Dante's inferno, drowning in agony screams.

In the morning, Dawn and our guide asked if I needed to rest for another day. I looked down at the sobbing woman, and said, "No, I'm good."

As bad as it was for me, it was worse for someone else. As long as I could limp, I would. I kept going and we got out of there on time and in one piece. I felt great about the experience, and now drag it out whenever I feel overwhelmed.

You have no choice but to be brave and keep going.

The Red Sea

I learned to scuba dive in Fiji. The first time I went under with the mask on was terrifying, an instant sensation of drowning. I panicked immediately and swam to the surface. I tried again, and luckily, I was distracted by my dive master's long snot booger that moved and bobbed underwater. I was laughing in my mask at how disgusting it was, and forgot all about drowning. Once I got over my initial fear, I swam with sea turtles and fish and fell in love. Scuba diving is addicting. Once you get a taste, you want to do it again and again, and go deeper each time. I've had the good fortune to dive in Australia along the Great Barrier Reef, in Bali and Thailand, and in the Red Sea in Dahab, Egypt.

While backpacking in Egypt, Lisa and I were on edge to begin with. To get from one side of the country to the other, we'd had to go with an armed caravan. Men with serious weapons were everywhere and you got a very real sense that you took your life into your own hands just being there. Two very American-looking young women, traveling through the desert, weren't exactly inconspicuous and we caught many looks that frightened and startled us.

I met this guy, a super-hot Parisian wearing a dive suit, on a beach. He was with a friend of his, a tall, sexy Dutchman. They told Lisa and me that they were scuba masters and asked if we'd like to go diving. We took one look at them, and glanced at each other, grinning. "Sure, we'd love to go diving with you," I said. They told us about a spot that was deeper than we were certified to go, but we lied and said we could do it.

"You have to swim through a four-foot-wide coral cave, straight down the whole way," said the French guy. "There are two openings into a lagoon, but if you miss the first one, you can't turn around because it's too narrow. You have to keep going to the second opening."

I just nodded, thinking, *I can do this*, perhaps crossing the mental line between bravery and stupidity.

So we plunged in, and swam headfirst into the tight tunnel, my tank scraping against the coral. The guys started gesturing at us, and I realized they were trying to say that we'd gone too far and missed the first opening. To get to the second, we'd have to descend much deeper than I'd ever been. While swimming the thirty feet to the second opening through the increasingly narrow passageway, I came very close to having a panic attack. As it got darker and darker, I thought, *I could die in this tunnel.*

Finally, the second opening appeared and we swam through it into a wide-open space with an incredible wall of fish and coral. We were probably the only four people for miles and it felt private and special. Lisa and I were blown away by the lush, jaw-dropping beauty, like a van Gogh or Kincaid painting. We followed the guys around until our tanks were nearly empty and we had to swim back through the tunnel and up to the surface.

Sharing these intense experiences—a death-defying dive and the dramatic reveal of a beautiful, hidden world—was romantic as hell. When the dive masters asked us to dinner that night, of course, we said yes. Lisa's Dutchman showed up in hipster glasses and a casual timeless style, and my Frenchman was shirtless in Thai elephant MC Hammer pants, and more yarn necklaces and bracelets than I was comfortable with. My date sure did present differently in civilian clothes; I will never forget the devilish grin Lisa gave me when we saw what he was wearing. In the end, I still made out with him, mostly just to get him out of those awful clothes. I didn't see him again after that, but he is locked in my memory forever. I'll never forget the sense of danger about Egypt itself, the dive, trusting strangers, and seeing the most thrilling natural beauty I've ever beheld.

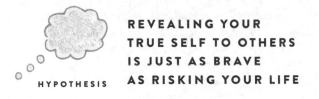

REVEALING YOUR TRUE SELF TO OTHERS IS JUST AS BRAVE AS RISKING YOUR LIFE

HYPOTHESIS

The Cut of Cheese

On *MythBusters*, we did an episode that answered the question, "Do pretty girls fart?" To test it, I was required to pass gas on camera. I think they threw in the word "pretty" to make me do it, thinking I'd be flattered, as in, *Oh, you think I'm pretty? I'll do whatever you want!*

They could have called me a goddess, I was still reluctant. I'd vomit on camera, sure. But fart? I had my dignity! Eventually, I agreed to take one for the team. It was just a little toot. I'd done worse.

They handed me a pair of what Jamie called "farting panties." (FYI: I can't stand it when fatherly men say the word "panties." It freaks me out. I made him call them "bloomers" or "underwear.") The knickers were outfitted with a hydrogen sulfide meter with an alarm and a microphone that was rigged up to a PA system for the entire

BRAVE HEARTS

As an artist and now an author, I draw inspiration from others. If you are ever in search of a brave woman to admire and emulate, explore the work and life stories of these artists who unfailingly turned their female hearts into their art:

Frida Kahlo. You know about her iconic flower-crowned self-portrait. What you might not know is that the Mexican artist/icon had polio as a child, causing her right leg to be shorter than her left, and that she survived a horrific bus accident at eighteen. Her pelvis was impaled by an iron rail that broke the bone, along with her collarbone, vertebrae, ribs, and both legs. She was in casts and bedridden for months, and lived with back pain and fatigue for the rest of her life. Many of her self-portraits explore her misshapen, damaged body, and what she endured. She also suffered through her two marriages to Diego Rivera, a passionate painter, who cheated on her constantly (as she cheated on him). Her later years were defined by health woes and heartache, but she pulled her canvases into bed, worked through her problems, and expressed her pain and passion in glorious, surrealist dreamlike pieces. For more about her life story, watch *Frida*, starring fellow Mexican artist Salma Hayek.

Simone de Beauvoir. The French author and philosopher wrote *The Second Sex* in 1949, and it's been the introduction to feminism for three generations of women, and counting. An activist and thinker, de Beauvoir pushed everything in her life through the filter of feminism, including her lifelong relationship with Jean-Paul Sartre. They never married. They weren't faithful. Their relationship didn't fit the traditional mold and no one was going to dictate to her what she could and couldn't do as a writer and a lover of men and women.

Anaïs Nin. French-born American Nin was a bohemian free spirit in Paris and New York, writing, seducing, living, and exploring emotions throughout the mid-twentieth century. She wrote deeply personal essays in her diaries, which she kept her whole life, and erotic story collections *Delta of Venus* and *Little Birds*. Her romantic life was as prolific as her literary output. Nin was married twice (both men at the same time for a while there), and took many lovers, most famously Henry Miller, who inspired her erotic writings. For me, reading her work in college was like watching a beautiful flower unfold. She wrote about fetish sex and bondage, "abnormal pleasures," themes I'd never heard about before, with such grace. She is credited as the first female erotica writer, defying convention and social dictates in her work and life. Although her name is not in the title, the movie *Henry & June* is about Nin's sexual awakening with Miller, and it's pretty hot.

Eva Hesse was one of the most influential postminimalist sculptors of the 1960s. Her brave use of experimental and unconventional materials like latex, fiberglass, and plastic created haunting sculptures that illuminated women's issues while refraining from any obvious political agenda. Her impulse to manipulate material but let it guide its own form inspires me today. Art was her heart. She was a juggernaut in a primarily male-dominated movement.

building. If I let one go, everybody would hear it. I was very tense, which I learned is a fart blocker. Adding to the stress, a bunch of executives from Discovery decided that they were going to tour the building *that day*. So it wasn't just like a few of my friends. There was a big audience standing around, waiting for me to release gas.

I had a wave of flashbacks to my coin bra high school moment, except this time, I was humiliating myself on purpose.

OWN IT

EXPERIMENT

They fed me cabbage and beans. I did jumping jacks. And nothing was happening in my high-tech undies. They were getting frustrated with me, and piling on the pressure to produce.

Adam was almost yelling at me. "You *have* to produce something!" And just then, in the final seconds of the show, I released a tiny "Chanel N° 2" emission. (We had to call it that because the network decided we couldn't use the word "fart" on TV.) Everybody cheered. I wanted to stand proud, but I just started giggling, and saying, "Oh my God, my parents are going to be watching this show. This is what I've become!"

Farting on TV was embarrassing, but even more, it was liberating.

Once you fart on camera, you own it. Everyone knows you do it. We *all* do it. That's why fart jokes are funny. I claimed the moment, and I knew everything was going to be okay. How could I ever be afraid to look (or sound) like a fool again?

Why was I *ever* afraid of that?

I had this idea in my head, that dignity meant hiding your humanity from the world. If you looked foolish, you lost your credibility. You don't see news anchors farting on camera, or even showing a zit. That would be too human, and, therefore, denigrating. Until that episode, I don't think I'd ever farted in front of my husband. I clung to the illusion that hiding the smelly, funny, icky parts of being human was how you put your best self forward. But everyone poops (as the book says). Everyone farts. Our audience enjoyed hearing me do it. When I freed myself of that so-called dignity, I became relatable on a new level. I was able to relax and be myself. I didn't even realize I'd been holding back a little until then.

My fart was revolutionary, the first in a shift toward human real-ness and grittiness on TV and the internet. Since then, we've seen the rise of #nofilter YouTube shows, and reality TV stars crying and look-ing like hell on camera. Being publicly unpolished is now acceptable, even desirable, because it's relatable.

Now freed from repression, I farted at will at the shop. After that, if Tory and Grant got in a fight on set, I'd ask one of them to pull my finger and then fart to break (wind and) the tension. A little gas was nothing to be afraid of.

ANALYSIS

BRAVERY IS SAYING "YES" EVEN IF YOU AREN'T SURE WHAT YOU'RE DOING

Head Rush

I was at a cocktail party for Discovery Channel and met Debbie Myers, the head of Science Channel at the time. I found her inspira-tional, a strong woman in a power position with incredible passion for her job and working for the hugely important cause of getting kids into science. She doesn't have a science background either, but learned the importance of it and how cool it is.

We talked about losing interest in science as a teen, and how a TV show for kids could reach that wavering audience and prevent it from

LITTLE SHOP
OF HORRORS

I did have terrifying moments on shows that still give me the chills. Besides the scorpion Sleeping Beauty test, these experiments also shook me to the timbers and tested my courage on *MythBusters* and beyond:

1. **Snakes.** We had a python on the set (maybe for the scorpion episode?) and I held it for a photo. It started winding around my neck and choking me. I was scared shitless, but I didn't want to admit it, so I just stood there while the snake throttled me, with glassy eyes and a smile. My tongue was dry and my whole body was tense, which was a signal to the snake, "Oh, good. Prey." Shortly after they took it off me, the snake bit the handler!

2. **Sharks.** For our Shark Week episodes, we continually came up with reasons to bait the water and swim around nose to nose with huge specimens, even at night a few times, in watery darkness, to see if they were attracted to flashlights! While filming in the Bahamas on our first shark trip, I was struck with how lucky I was to be on the show and to dive with sharks on a random Tuesday. Overwhelmed with gratitude, I went up to Jamie who was relaxing on a lounge chair, and started tearing up, okay maybe blubbering a bit, "I just want to thank you for this. I can't believe this is my job and that I'm here." I basically threw up my emotions on a guy who just didn't go there, ever. He looked at me like he was touched, but more uncomfortable. He said something like, "Good having you here" and escaped. I still have a real soft spot for Jamie for giving me my first chance.

3. **Chinese water torture.** I hated this one. Early days, we tested dropping water on your head for a long period of time. I said, "What's the goal here? To prove torture?" I was going to be a good sport about it, and let them tie me down, which I also thought was not good for the experiment. It added another variable, and sure enough, my arms started cramping. I thought about climbing the Himalayas, and getting through it, but I wound up just losing it. It *was* torture! I lasted forty-five minutes. I had this moment of dread and thought, *We are messing with some dark shit. Do we really want to put this on the air?* We brought in a guy who had been tortured as an expert, and he said, "What are you guys doing?" I wish we didn't air it, but we did.

4. **Adrenaline stunt.** In *Thrill Factor*, we did an episode in Las Vegas and went on a ride that was a decelerated jump off a hundred-story building. The experience was only a couple of minutes but I built up a ton of anxiety the whole week before the shoot. I would be sleeping in my bed and wake up with a jolt about jumping off a building. I had to remind myself, *You're safe in your bed. It's okay.* I'd be putting on my makeup, think about it, and my heart would start hammering away. I was more afraid before the leap, by far, for far longer, than I was doing it. I learned a valuable lesson about anticipatory dread, aka anxiety: Don't worry until you have something to worry about in the moment. For the actual jump, I was smiling and giggling (what I do when nervous), so it looked like I could handle it. I wasn't thinking about being brave as I stepped up to the edge. I reminded myself that I'm just a lifelong student, learning every day and every way how not to let fear stop me. This was just one more challenge, in a long series of them, to fling myself into the unknown. So I did.

happening, and build interest in their parents, too, so they could be reawakened to the wonder.

Debbie and I started talking seriously about how we could create a show—hosted by me alone—that would achieve that goal and *Head Rush* was born. It was a half-hour DIY show for kids on the Science Channel, very much a passion project.

Until then, I'd only been part of a collaborative team. Jamie and Adam helmed *MythBusters*. The success rested on them. *Head Rush* was the first time I was in the driver's seat and would carry the entire show. On top of that, I'd set out to create something of educational value at the outset (*MythBusters* did that but as a happy side effect). *Head Rush* had a lofty purpose and with that came a heavy weight on my shoulders. Stepping out of the background into the spotlight as the show's one and only anchor, with a grandiose objective of trying to improve the world at large, was an act of bravery unlike any I'd made before.

I brought a production team from *MythBusters* with me. We had to work within a very small budget. The network came up with what they thought they could do, a scripted clip show with a little hosting. I went back to my team, camera and sound guys who worked Saturdays for day rates, and my genius producer/director Lauren Williams, and said, "I want more pizzazz."

Lauren agreed, and said, "They asked for a VW. Let's give them a Porsche."

We pulled out all the tricks, cool lighting, dramatic (yet cheap) experiments, a modern filming style. Everyone brought their A games and pitched in. I wore brightly colored T-shirts to make myself pop and keep the aesthetic interesting for kids at home. We pulled in people to talk about how they used science in their world, for example, Tony Hawk showed us how he uses math to design skateboard ramps.

I'd just given birth and was breastfeeding while we made this show; my schedule was unforgiving. Pump, drive Stella and the bottles to my parents' house, drive back to shoot on a Saturday, drive back to Los Gatos to pick up the kid, back to SF. And then, back to work at *MythBusters* all week. I was exhausted, but I *had* to do this show. I had this idea of going back in time, with my younger self seeing it, then maybe going to engineering school. As hard as it was, *Head Rush* was worth it to create my own show. **I got a taste of being the boss of my own vision.**

We did two seasons. Unfortunately, we didn't reach the right audience. Thirteen-year-olds don't watch the Science Channel. Kids don't watch TV, they watch their computers. Fittingly, *Head Rush* continues to have a second life in science classrooms all over America, thanks to the internet. If I produced my own show again for this audience, I'd do it online.

Through *Head Rush* and the connections I made at Science Channel, I was invited to the White House Science Fair (the Obama administration actually contributed funding for the show). It also took me to SXSW, and around the world to promote science. I did shows in Poland and Russia, and saw a new part of the world—all because of a brave first step I took at a cocktail party.

CRASH AND LEARN

CONCLUSION

"Be brave" has been my life mission statement in general, starting when I was that scared shy girl. It was obvious to me even then that courage in the face of fear is the main ingredient in any recipe for success and happiness. I pushed myself to do one scary thing or another, just to see what would happen. Whether I crashed and burned or walked away triumphant didn't matter as much as being able to put myself in the fray in the first place. In other words, process was more important than outcome, because regardless of how things played out, I got a little bit stronger and wiser each time.

By being covered in scorpions, or wearing the coin bra, or diving though a coral tunnel, or farting on camera, for example, I came to the obvious conclusion that **bravery is being afraid of something and doing it anyway.**

Bravery isn't being a superhero. It's what you have to do to be free.

Bravery is not giving a shit.

It's getting messy and facing the consequences.

It's being who you are, and where you are, right now.

When I was at the White House Science Fair, my vision of confidence and bravery was redefined at the roundtable discussion later with special advisor Valerie Jarrett, Michelle Obama's chief of staff Tina Tchen, Ellen Stofan of NASA, and some award-winning, inspirational student scientists. We went around the table and talked about what interests and experiences brought us all here today. So many inspiring stories of science love and overcoming odds from fourteen-year-old kids! Eventually, the question, "What brought you here today?" worked around the table to me. I was humbled to be at this table, but stuck to my bedrock belief, that bravery is the absence of dignity and the ability to laugh at yourself. So, to the luminaries and geniuses, I said, "I'm here because I like to blow stuff up."

My life's mission wasn't to be famous or on TV, and definitely not to work in science. I just wanted creativity, passion, and adventure. Who knew that my life would bring my dreams to me? I've stumbled into my career, my marriage, everything, and have fought like hell to grab every opportunity, including writing this book and getting this message to you.

You can't wait for the fancy invitation, the golden door with a neon sign that says, YOUR LIFE THIS WAY, or the giant X on the treasure map. You'll be waiting for eternity. You have to choose your own adventure, take the trip, enroll in the class, say "yes" when you are afraid. You are the X on the map.

I started out a Crash Test Girl, when I was very much a little kid. It took me a long time to be a Road-Tested Woman. From where I stand, I look back and appreciate how much I had to learn to get to where I am now, and that at every step of the way, I was growing and evolving, through trial after trial by fire. No wonder I love blowing things up on TV and in life. It's all about the explosion—of flames, of insight, of ideas—the chemical reactions I create through practical, methodical experimentation.

Sometimes, I crash and burn. Always, I crash and learn. By using "girl," I remind myself to keep the wonder and fearlessness of a child as I continue to blow shit up as a grown-ass woman. My lesson for bravery—the one lesson to rule them all—is Crash and Learn (shades, boots, and burning car in the background optional).

DIY

CRASH
TEST

Now it's your turn.

Use the scientific method to solve

your own toughest questions...

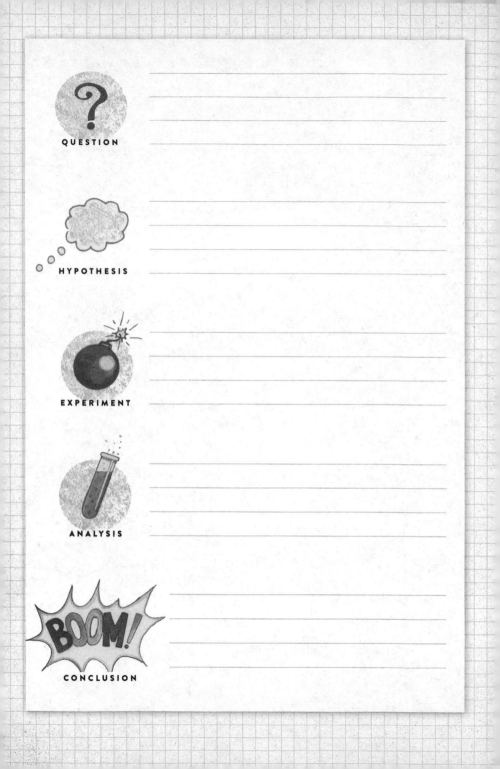

QUESTION

HYPOTHESIS

EXPERIMENT

ANALYSIS

BOOM!

CONCLUSION

QUESTION

HYPOTHESIS

EXPERIMENT

ANALYSIS

BOOM!

CONCLUSION

QUESTION

HYPOTHESIS

EXPERIMENT

ANALYSIS

BOOM!

CONCLUSION

QUESTION

HYPOTHESIS

EXPERIMENT

ANALYSIS

CONCLUSION

ACKNOWLEDGMENTS

- - - - - - -

Thank you, Sue and Dennis Byron, for letting the wild flower grow in your garden. Thank you, little crash test sister, Summer Youdall, for holding my hand and sharing the crazy, remembering every detail of our childhood. Thanks to the rebel yell that became science-based reality television and the chance Discovery took on a mythbusting girl. Jamie Hyneman gave me the ticket to this thrill ride. The *MythBusters* gang and all the crew/family rode that coaster through the turns, climbs, drops, and sometimes inversions (especially Tory Belleci, who keeps running back to the front of the line with me). Thank you, Linda Wolkovitch, Lauren Williams, Yvette Solis, Francesca Garigue, Jacelyn Maker, Jaime Garcia, and Kristen Lomasney, the wild "ride or die" women of *MythBusters* who remain my sounding board and have my back at all times. Thanks to the bomb techs, sheriffs, police, firefighters, stunt men, range masters, scientists, and former FBI agents who kept me safe and had the best drinking stories I have ever heard and will never repeat.

Lisa Cole, you will always be a star in my sky and the best words in my story. Brittany Smail, for a lifelong friendship and helping me breathe under the weight of trying to write a book. Of course, Homer Hickman for the last-call double-dog-dare to write it, and Frank Weiman, Val Frankel, Hilary Lawson, Sydney Rogers, and Janet Evans-Scanlon for making it really happen. I hope within this book my daughter, Stella, will someday find a love note to the amazing beautiful woman she will become. And a love note to the man who knows all my demons and still invited them in to play. I love you, Paul Urich. Thank you. Finally, thank you to all the fans that continue to support me through every insane project that I put out into the world. The reason I have this incredible life is because you decided to watch it.

ABOUT THE AUTHOR

- - - - - - -

KARI BYRON has been the most recognizable, honored, and beloved woman in science-based reality TV for over a decade. She is best known as a host on Discovery Channel's *MythBusters* but has gone on to host and produce shows spanning several networks: *Head Rush*; *Punkin Chunkin*; *Large, Dangerous Rocket Ships*; *Thrill Factor*; *Strange Trips*; *America Declassified*; *White Rabbit Project*; and *Positive Energy*.

Fostering her curious nature, she spent a lifetime acquiring odd skills and interests. After graduating from college and traveling the world, she decided to settle in San Francisco and pursue a career as a sculptor. Special effects and prop making seemed like a perfect fit. Trying to break into the field, she got an internship at M5 Industries with Jamie Hyneman. Her first day turned out to be the beginning of *MythBusters* and a career in television. Though her dream was to be a working artist, in a strange turn of events, her real dream job turned out to involve handling poop, eating bugs, wiring explosives, and scraping chicken guts off the ceiling.

Her passion for smart entertainment has led her around the world, speaking about the role of STEAM (science, technology, engineering, art, and mathematics) programming on television and touring with a live show, as well as hosting science fairs from Google's headquarters to the White House. Her new adventure is creating global citizens through cultural understanding, with Crash Test World streaming internationally.

Currently she lives in San Francisco but as a global citizen, you never know where in the world you may find her.